SECURITY OF RADIOACTIVE MATERIAL IN USE AND STORAGE AND OF ASSOCIATED FACILITIES

The following States are Members of the International Atomic Energy Agency:

AFGHANISTAN	GERMANY	PAKISTAN
ALBANIA	GHANA	PALAU
ALGERIA	GREECE	PANAMA
ANGOLA	GRENADA	PAPUA NEW GUINEA
ANTIGUA AND BARBUDA	GUATEMALA	PARAGUAY
ARGENTINA	GUYANA	PERU
ARMENIA	HAITI	PHILIPPINES
AUSTRALIA	HOLY SEE	POLAND
AUSTRIA	HONDURAS	PORTUGAL
AZERBAIJAN	HUNGARY	QATAR
BAHAMAS	ICELAND	REPUBLIC OF MOLDOVA
BAHRAIN	INDIA	ROMANIA
BANGLADESH	INDONESIA	RUSSIAN FEDERATION
BARBADOS	IRAN, ISLAMIC REPUBLIC OF	RWANDA
BELARUS	IRAQ	SAINT LUCIA
BELGIUM	IRELAND	SAINT VINCENT AND
BELIZE	ISRAEL	THE GRENADINES
BENIN	ITALY	SAN MARINO
BOLIVIA, PLURINATIONAL	JAMAICA	SAUDI ARABIA
STATE OF	JAPAN	SENEGAL
BOSNIA AND HERZEGOVINA	JORDAN	SERBIA
BOTSWANA	KAZAKHSTAN	SEYCHELLES
BRAZIL	KENYA	SIERRA LEONE
BRUNEI DARUSSALAM	KOREA, REPUBLIC OF	SINGAPORE
BULGARIA	KUWAIT	SLOVAKIA
BURKINA FASO	KYRGYZSTAN	SLOVENIA
BURUNDI	LAO PEOPLE'S DEMOCRATIC	SOUTH AFRICA
CAMBODIA	REPUBLIC	SPAIN
CAMEROON	LATVIA	SRI LANKA
CANADA	LEBANON	SUDAN
CENTRAL AFRICAN	LESOTHO	SWEDEN
REPUBLIC	LIBERIA	SWITZERLAND
CHAD	LIBYA	SYRIAN ARAB REPUBLIC
CHILE	LIECHTENSTEIN	TAJIKISTAN
CHINA	LITHUANIA	THAILAND
COLOMBIA	LUXEMBOURG	TOGO
CONGO	MADAGASCAR	TRINIDAD AND TOBAGO
COSTA RICA	MALAWI	TUNISIA
CÔTE D'IVOIRE	MALAYSIA	TURKEY
CROATIA	MALI	TURKMENISTAN
CUBA	MALTA	UGANDA
CYPRUS	MARSHALL ISLANDS	UKRAINE
CZECH REPUBLIC	MAURITANIA	UNITED ARAB EMIRATES
DEMOCRATIC REPUBLIC	MAURITIUS	UNITED KINGDOM OF
OF THE CONGO	MEXICO	GREAT BRITAIN AND
DENMARK	MONACO	NORTHERN IRELAND
DJIBOUTI	MONGOLIA	UNITED REPUBLIC
DOMINICA	MONTENEGRO	OF TANZANIA
DOMINICAN REPUBLIC	MOROCCO	UNITED STATES OF AMERICA
ECUADOR	MOZAMBIQUE	URUGUAY
EGYPT	MYANMAR	UZBEKISTAN
EL SALVADOR	NAMIBIA	VANUATU
ERITREA	NEPAL	VENEZUELA, BOLIVARIAN
ESTONIA	NETHERLANDS	REPUBLIC OF
ESWATINI	NEW ZEALAND	VIET NAM
ETHIOPIA	NICARAGUA	YEMEN
FIJI	NIGER	ZAMBIA
FINLAND	NIGERIA	ZIMBABWE
FRANCE	NORTH MACEDONIA	
GABON	NORWAY	
GEORGIA	OMAN	

The Agency's Statute was approved on 23 October 1956 by the Conference on the Statute of the IAEA held at United Nations Headquarters, New York; it entered into force on 29 July 1957. The Headquarters of the Agency are situated in Vienna. Its principal objective is "to accelerate and enlarge the contribution of atomic energy to peace, health and prosperity throughout the world".

IAEA NUCLEAR SECURITY SERIES No. 11-G (Rev. 1)

SECURITY OF RADIOACTIVE MATERIAL IN USE AND STORAGE AND OF ASSOCIATED FACILITIES

IMPLEMENTING GUIDE

INTERNATIONAL ATOMIC ENERGY AGENCY
VIENNA, 2019

IAEA Library Cataloguing in Publication Data

Names: International Atomic Energy Agency.
Title: Security of radioactive material in use and storage and of associated facilities / International Atomic Energy Agency.
Description: Vienna : International Atomic Energy Agency, 2019. | Series: IAEA nuclear security series, ISSN 1816–9317 ; no. 11-G (Rev. 1) | Includes bibliographical references.
Identifiers: IAEAL 19–01263 | ISBN 978–92–0–110018–4 (paperback : alk. paper)
Subjects: LCSH: Radioactive substances. | Nuclear industry — Security measures. | Nuclear facilities.
Classification: UDC 620.267:343.852 | STI/PUB/1840

FOREWORD

The IAEA's principal objective under its Statute is "to accelerate and enlarge the contribution of atomic energy to peace, health and prosperity throughout the world." Our work involves both preventing the spread of nuclear weapons and ensuring that nuclear technology is made available for peaceful purposes in areas such as health and agriculture. It is essential that all nuclear and other radioactive materials, and the facilities at which they are held, are managed in a safe manner and properly protected against criminal or intentional unauthorized acts.

Nuclear security is the responsibility of each individual State, but international cooperation is vital to support States in establishing and maintaining effective nuclear security regimes. The central role of the IAEA in facilitating such cooperation and providing assistance to States is well recognized. The IAEA's role reflects its broad membership, its mandate, its unique expertise and its long experience of providing technical assistance and specialist, practical guidance to States.

Since 2006, the IAEA has issued Nuclear Security Series publications to help States to establish effective national nuclear security regimes. These publications complement international legal instruments on nuclear security, such as the Convention on the Physical Protection of Nuclear Material and its Amendment, the International Convention for the Suppression of Acts of Nuclear Terrorism, United Nations Security Council resolutions 1373 and 1540, and the Code of Conduct on the Safety and Security of Radioactive Sources.

Guidance is developed with the active involvement of experts from IAEA Member States, which ensures that it reflects a consensus on good practices in nuclear security. The IAEA Nuclear Security Guidance Committee, established in March 2012 and made up of Member States' representatives, reviews and approves draft publications in the Nuclear Security Series as they are developed.

The IAEA will continue to work with its Member States to ensure that the benefits of peaceful nuclear technology are made available to improve the health, well-being and prosperity of people worldwide.

CONTENTS

1. INTRODUCTION

1.1. The IAEA Nuclear Security Series provides guidance to Member States to assist them in implementing a national nuclear security regime, and in reviewing and, when necessary, strengthening this regime. The series also provides guidance to States in fulfilling their obligations and commitments with respect to binding and non-binding international instruments. The Nuclear Security Fundamentals set out the objective of a nuclear security regime and its essential elements [1]. The following publications indicate what a nuclear security regime should address:

— Nuclear Security Recommendations on Physical Protection of Nuclear Material and Nuclear Facilities [2];
— Nuclear Security Recommendations on Radioactive Material and Associated Facilities [3];
— Nuclear Security Recommendations on Nuclear and Other Radioactive Material out of Regulatory Control [4].

This publication is the primary Implementing Guide for the Nuclear Security Recommendations on Radioactive Material and Associated Facilities [3].

1.2. This Implementing Guide is a revision of IAEA Nuclear Security Series No. 11, Security of Radioactive Sources, published in 2009. This revision was undertaken to:

(a) Better align this publication with the recommendations contained in Ref. [3], first published in 2011;
(b) Expand the scope of the guidance to include not only radioactive sources as defined in the Code of Conduct on the Safety and Security of Radioactive Sources [5], but also to address all radioactive material and associated facilities as defined in Ref. [3];
(c) Cross-reference other relevant guidance published since 2009;
(d) Add detail on selected topics based on the experience of the IAEA and Member States in using the previous version of IAEA Nuclear Security Series No. 11.

OBJECTIVE

1.3. The objective of this publication is to provide guidance to States and their competent authorities on how to establish or improve, implement, maintain and sustain the elements of the nuclear security regime related to radioactive material, associated facilities and associated activities, with particular emphasis on the development of regulatory requirements.

1.4. The present publication provides guidance to States in implementing the elements of a nuclear security regime related to radioactive material, including potential obligations and commitments with respect to relevant international instruments, such as the International Convention for the Suppression of Acts of Nuclear Terrorism [6], the Code of Conduct on the Safety and Security of Radioactive Sources [5] and its supplementary Guidance on the Management of Disused Radioactive Sources [7] and Guidance on the Import and Export of Radioactive Sources [8].

1.5. Many States have applied the guidance provided in the 2009 version of the Implementing Guide in establishing regulatory requirements for the security of radioactive sources. The publication of this revised version is not intended to be interpreted as advising that States need to amend their regulations to be consistent with the revised guidance, for example to address the security of radioactive material other than radioactive sources. However, they may choose to expand the scope of their regulatory programmes or make modifications over time, consistent with national priorities and changing circumstances, such as threat.

SCOPE

1.6. This publication applies to the security of radioactive material in use or in storage, as well as associated facilities and associated activities, against unauthorized removal of the radioactive material and sabotage performed with the intent to cause harmful radiological consequences. In this publication, security refers to both security systems and security management measures.

1.7. This publication addresses the security of radioactive material throughout its life cycle, including manufacture, supply, receipt, possession, storage, use, transfer, import, export, maintenance, recycling and disposal.

1.8. As used in this publication, radioactive material includes radioactive sources and unsealed radioactive material under regulatory control, including radioactive material over which regulatory control has been gained or regained. Where appropriate, States may also consider the application of this guidance to radioactive waste. The term 'radioactive material' is used throughout this guide, but the application of this guidance to radioactive material other than radioactive sources will depend on national context and priorities.

1.9. While this publication applies to protection against both unauthorized removal and sabotage, the detailed guidance primarily addresses measures to protect against unauthorized removal. These measures will also provide some capability to counter sabotage. However, to the extent that sabotage represents a particular concern to the State or the regulatory body, additional or more stringent security measures beyond those discussed in this guidance may be appropriate.

1.10. This publication does not cover preparedness and response to a nuclear or radiological emergency triggered by a nuclear security event, which are addressed in Refs [9, 10].

1.11. This publication also does not provide detailed guidance on security of radioactive material in transport, which is addressed in specific guidance [11].

1.12. This publication does not apply to the physical protection of nuclear material against unauthorized removal for use in a nuclear explosive device or to the physical protection of nuclear facilities against sabotage. These topics are addressed in Ref. [2] and its supporting Implementing Guide [12]. When a facility contains nuclear material and other radioactive material, the protection requirements for both should be considered and implemented in a consistent and non-conflicting manner in order to achieve an adequate level of security.

1.13. This publication assumes that the State has established and implemented a legislative and regulatory framework for the control and safety of radioactive material and associated facilities, including enabling legislation, a regulatory body, a national register (inventory) of radioactive sources, an authorization process, regulatory requirements for safety and provisions for inspection and enforcement. In this publication, the term 'protection and safety' is intended to include radiation protection. Such elements are addressed more completely in Refs [5, 13–16].

STRUCTURE

1.14. Following this introduction, Section 2 sets out the objectives of the elements of a State's nuclear security regime related to radioactive material, associated facilities and associated activities. Section 3 provides guidance to States and their competent authorities on the elements of the States' nuclear security regimes related to radioactive material, associated facilities and activities.[1] Section 4 provides guidance on key security concepts related to the security of radioactive material. Sections 5 and 6 expand on the guidance provided in Sections 2–4, focusing on the establishment of a State's regulatory programme for radioactive material. Section 5 provides guidance regarding the development of regulatory requirements for the security of radioactive material. Section 6 gives detailed guidance on establishing regulatory requirements using a prescriptive approach and more general guidance on performance based and combined approaches. Three appendices provide a description of security measures discussed in this guide (Appendix I); an outline of topics to be addressed in an operator's security plan (Appendix II); and a description of a vulnerability assessment (Appendix III).

2. OBJECTIVES OF A STATE'S NUCLEAR SECURITY REGIME RELATED TO RADIOACTIVE MATERIAL, ASSOCIATED FACILITIES AND ASSOCIATED ACTIVITIES

2.1. According to para. 2.1 of Ref. [1], "The objective of a State's *nuclear security regime* is to protect persons, property, society, and the environment from harmful consequences of a *nuclear security event*."

2.2. Malicious acts involving radioactive material, associated facilities and associated activities that could result in a nuclear security event include:

— Unauthorized removal of radioactive material for:
  • Use in a radiological dispersal device, a device designed to spread radioactive material using conventional explosives, or by other means, for the purpose of causing health effects or contaminating ground, buildings and infrastructure, leading to denial of access to these areas, or denial of service from the infrastructure;
  • Use in a radiation exposure device, a device designed to intentionally expose members of the public to radiation, such as the deliberate

---

[1] Sections 2 and 3 of this publication approximately follow the structure of the related Nuclear Security Recommendations publication [3].

4

placement of unshielded radioactive material in a public area, or the
deliberate placement of radioactive material in food or water to cause
radiation doses or poisoning through ingestion.
— Sabotage of radioactive material or an associated facility order to achieve
one or more of the same purposes.

2.3. According to para. 2.1 of Ref. [3], the objectives of a nuclear security regime
for radioactive material, associated facilities and associated activities should be:

"— Protection against *unauthorized removal* of *radioactive material* used
in *associated facilities* and in *associated activities*;
— Protection against *sabotage* of *other radioactive material, associated
facilities* and *associated activities*;
— Ensuring the implementation of rapid and comprehensive measures
to locate, recover, as appropriate, *radioactive material* which is lost,
missing or stolen and to re-establish regulatory control."

MEANS OF ACHIEVING THE OBJECTIVES

2.4. Paragraph 2.2 of Ref [3] states: "These objectives are realized through
security measures to deter, detect, delay and respond to a potential *malicious act*,
and to provide for the security management of *radioactive material* and *associated
facilities* and *associated activities*."

2.5. "These security measures should be based on a risk informed *graded
approach*" [3], taking into account the principles of risk management, including
such considerations as the potential radiological consequences of a malicious
act, the level of threat and the relative attractiveness of the radioactive material
for a malicious act (based on such factors as quantity, physical and chemical
properties, mobility, availability and accessibility). Appropriate security measures
should be adapted depending on whether the radioactive material concerned is
sealed, unsealed, disused or waste. This graded approach ensures that the highest
consequence material receives the greatest degree of security.

2.6. Paragraph 2.4 of Ref [3] states: "Recognizing the societal benefits of using
radioactive material, the *nuclear security regime* should strive to achieve a
balance between managing *radioactive material* securely without unduly limiting
the conduct of those beneficial activities."

3. ELEMENTS OF A STATE'S NUCLEAR SECURITY REGIME RELATED TO RADIOACTIVE MATERIAL, ASSOCIATED FACILITIES AND ASSOCIATED ACTIVITIES

3.1. This section provides guidance on the principles, concepts and approaches for implementing the elements of the nuclear security regime related to radioactive material, associated facilities and associated activities, based on the recommendations contained in Ref. [3].

STATE RESPONSIBILITY

3.2. Paragraph 3.1 of Ref. [3] states: "The responsibility for the establishment, implementation and maintenance of a *nuclear security regime* within a State rests entirely with that State."

3.3. The State[2] should take appropriate steps to ensure that the nuclear security regime encompasses the protection of radioactive material within the State's territory, or under its jurisdiction or control. The operator should be assigned prime responsibility for implementing and maintaining security measures for radioactive material, associated facilities and associated activities.

ASSIGNMENT OF NUCLEAR SECURITY RESPONSIBILITIES

3.4. Paragraph 3.2 of Ref. [3] states:

"The State should clearly define and assign nuclear security responsibilities to *competent authorities*, noting that they may include *regulatory bodies*, law enforcement, customs and border control, intelligence and security agencies, health agencies, etc."

---

[2] In some sections of this publication the distinction between the State and its competent authorities has not been precisely defined. This ambiguity recognizes the differences among States with respect to the assignment of responsibilities among a State's competent authorities. Nonetheless, a State should be specific and comprehensive in assigning and documenting nuclear security responsibilities.

3.5. The State should clearly define and assign nuclear security responsibilities to one or more competent authorities and confer upon each the powers necessary to perform their assigned functions. Table 1 depicts a typical assignment of nuclear security responsibilities to competent authorities. A State's actual assignment of such responsibilities may vary depending on national law, practice and circumstances. However, each responsibility indicated in the second column of Table 1 should be assigned to at least one competent authority.

3.6. Paragraphs 3.2 and 3.3 of Ref. [3] state respectively:

"Provision should be made for appropriate integration and coordination of responsibilities within the State's *nuclear security regime*. Clear lines of responsibility and communication should be established and recorded between the *competent authorities*."

"The State should ensure effective overall cooperation and relevant information sharing between the *competent authorities*. This should include sharing of relevant information (such as information about the *threat* to be protected against and other useful intelligence) in accordance with national regulations."

3.7. The State may consider establishing a coordinating body that includes representatives of competent authorities with assigned nuclear security responsibilities and that meets regularly for the purpose of ensuring adequate integration, communication and coordination. One of the competent authorities should be assigned as the lead of the coordinating body. The State may consider promoting the use of such instruments as memoranda of understanding, inter-agency agreements and the like as a means of facilitating cooperation and information sharing among competent authorities.

TABLE 1.   TYPICAL ASSIGNMENT OF NUCLEAR SECURITY RESPONSIBILITIES

Competent authority	Nuclear security responsibilities and powers
Regulatory body	Establish a system of regulatory control over radioactive material, associated facilities and associated activities that places the primary responsibility for nuclear security on authorized persons (licensees)
	Establish a system of security based categorization
	Develop and maintain a national register of radioactive material over activity thresholds defined by the State
	Participate in national threat assessment
	Develop and apply the design basis threat, representative threat statement or other defined threat for purposes of regulation for security
	Implement the authorization (licensing) process, including review and assessment of security systems and security management measures
	Establish regulatory requirements and provide guidelines for security, including requirements for information protection
	Manage the safety–security interface
	Conduct security inspections
	Take enforcement action for non-compliance
	Participate in regional and international databases and other cooperative activities
	Encourage and promote a robust nuclear security culture
	Participate in planning efforts for preparedness for and response to nuclear security events, including exercises
	Administer procedures for authorizing and controlling the import and export of radioactive material
	Notify operators concerning specific or increased threat
	Review and assess the design of security systems (in the authorization process)
Law enforcement	Provide response to interrupt malicious acts (e.g. unauthorized access, unauthorized removal, sabotage)
	Participate in planning efforts for preparedness for and response to nuclear security events, including exercises
	Participate in national threat assessment
	Identify facility or activity specific threats, or new or increased threat capabilities
	Conduct background checks for purposes of trustworthiness verification
	Detect and investigate nuclear security events

TABLE 1.   TYPICAL ASSIGNMENT OF NUCLEAR SECURITY RESPONSIBILITIES (cont.)

Competent authority	Nuclear security responsibilities and powers
Customs and border control	Participate in national threat assessment Identify facility or activity specific threats, or new or increased threat capabilities Control and detect non-compliance with respect to imports or exports Communicate with regulatory body with respect to national inventory of radioactive material
Intelligence and security agencies	Direct national threat assessment Identify specific or increased threats
National emergency response agency	Coordinate planning efforts for preparedness for and response to nuclear security events
Civil defence, health and environment agencies	Participate in planning efforts for preparedness for and response to nuclear security events
Ministry of justice and prosecuting authorities	Prosecute alleged perpetrators of malicious acts
Ministry of foreign affairs	Engage in regional and international cooperation

LEGISLATIVE AND REGULATORY FRAMEWORK

State

3.8.   Paragraph 3.4 of Ref. [3] states:

"The State should establish, implement, and maintain an effective national legislative and regulatory framework to regulate the nuclear security of *radioactive material, associated facilities* and *associated activities*, which:

— Takes into account the risk of *malicious acts* involving *radioactive material* that could cause *unacceptable radiological consequences*;

— Defines the *radioactive material, associated facilities* and *associated activities* which are subject to the *nuclear security regime* in terms of nuclides and quantities of *radioactive material* present;
— Prescribes and assigns governmental responsibilities to relevant entities including an independent *regulatory body*;
— Places the prime responsibility on the *operator* … for implementing and maintaining security measures for *radioactive material*;
— Establishes the *authorization* process for *radioactive material, associated facilities* and *associated activities*. As appropriate, the *authorization* process concerning the security of *radioactive material* could be integrated within one defined for safety or radiation protection;
— Establishes the inspection process for security requirements;
— Establishes the enforcement process for the failure to comply with security requirements established under [the] legislative and regulatory framework;
— Establishes sanctions against the *unauthorized removal* of *radioactive material* and *sabotage* of *associated facilities* and *associated activities*;
— Takes into account the interface between security and safety of *radioactive material*."

3.9. As indicated in Refs [13, 17], in most States, the legal hierarchy consists of several levels: constitutional instruments; statutory instruments, also referred to as primary legislation[3]; regulations; and non-mandatory guidance instruments such as agreements among competent authorities and associated administrative measures.[4] Table 2 depicts topics addressed by an example legislative and regulatory framework for radioactive material, associated facilities and associated activities based on the State's legal hierarchy. This example is intended to provide a starting point for establishing or strengthening such a framework. Depending on the specifics of the State's legal hierarchy, the level at which these topics are addressed may vary and the detailed contents of a State's legislative and regulatory framework should reflect its national practice and needs.

3.10. The legislative and regulatory framework for the security of radioactive material should take into account the legislative and regulatory framework for radiation protection and safety. Often, a single regulatory body is responsible for authorization and oversight for both safety and security, in which case

---

[3] Primary legislation refers to enactments by a parliament or other legislature.

[4] A fuller description of a State's legal hierarchy is included in Ref. [16], including further discussion of the first level of the legal hierarchy — constitutional instruments — which is outside the scope of the current publication.

TABLE 2.   AN EXAMPLE LEGISLATIVE AND REGULATORY FRAMEWORK

Level	Topics addressed
Primary legislation	Definition of radioactive material, associated facilities and associated activities subject to regulatory control with respect to nuclear security Establishment or designation of competent authorities with defined responsibilities and powers in relation to nuclear security Definition of offences and establishment of penalties related to nuclear security Establishment of security goals and sub-goals
Regulations	Authorization (licensing) process Security requirements, including requirements for information security Import–export requirements Requirements regarding transfer of radioactive material Requirements regarding inventory taking and reporting to national register Inspection and enforcement process
Agreements	Agreement among competent authorities regarding exchange of threat information Agreement between regulatory body and law enforcement regarding the conduct of background checks Agreement between regulatory body and law enforcement regarding response to interrupt malicious acts Agreement between regulatory body and ministry of justice or other prosecuting authority regarding referrals for prosecution Coordination agreement between safety regulatory body and nuclear security regulatory body (if separate) Coordination agreement between regulatory bodies with separate practice based jurisdictions respecting security of radioactive material (for example, industrial versus health care)
Associated administrative measures	Authorization (licensing) procedures and forms Guidance on implementation of security requirements, including guidance on verifying trustworthiness, safety–security interface Model security plan Radioactive material inventory and reporting forms Security inspection manual, including forms Enforcement policy

authorization could be conducted in a single, integrated process. If the same regulatory body is not responsible for both safety and security, there should be regular, systematic cooperation and information sharing between the regulatory bodies for safety and security. Regardless of the organization of the regulatory system in place, the interfaces between safety and security requirements should be appropriately managed.

3.11. Paragraph 3.5 of Ref. [3] states:

"The State should take appropriate steps within the legislative and regulatory framework to establish and ensure the proper implementation of its *nuclear security regime* throughout the life cycle of the *radioactive material*."

3.12. The competent authority should require an authorization for activities involving radioactive material above a certain activity threshold as defined by the State. The competent authority should regulate all activities involving this radioactive material for security purposes, from manufacture through supply, receipt, possession, storage, use, transfer, import, export, maintenance, recycling and disposal.

3.13. In many States, there is a single authorization addressing both safety and security for any activity involving radioactive material. In order to obtain the authorization, the State should require that the applicant demonstrate its ability to meet applicable safety and security requirements. Once the applicant has made this demonstration and the regulatory body has issued the authorization, continued compliance with the applicable safety and security requirements is typically made a condition of the authorization. In some States, an authorization for safety purposes may be already in place when the regulatory body establishes security requirements for radioactive material, associated facilities and associated activities. In such cases, the regulatory body should ensure that these security requirements are mandatory for existing authorization holders, for example through amendment of existing authorizations or by including in the security requirements a specific provision that they are mandatory for existing authorization holders.

3.14. Paragraph 3.6 of Ref. [3] states:

"The State should designate one or more *competent authorities*, including a *regulatory body*, for the establishment, implementation and maintenance of a *nuclear security regime*, which have a clearly defined legal status and independence from the *operator*... and which have the legal authority to enable them to perform their responsibilities and functions effectively."

3.15. The State could choose to designate a single regulatory body with responsibility for authorization, inspection and enforcement of the security of all radioactive material, or designate more than one such body with shared responsibilities and functions that depend on how the radioactive material will be used. For example, one regulatory body may hold jurisdiction over medical uses of radioactive material and another jurisdiction over industrial and other uses. In such cases, the boundary between the respective jurisdictions should be clearly drawn and the regulatory approaches should be consistent and compatible.

3.16. Regardless of the approach taken, the regulatory body should be independent of the operators which they regulate. Practices that promote such independence and which should be considered include:

— Functional separation of the regulatory body from entities having responsibilities or interests that could unduly influence decision making.
— Refraining from assigning responsibilities to the regulatory body that might compromise or conflict with the discharge of its responsibilities for regulating the security of facilities and activities.
— Prohibition of direct or indirect interest of the staff of the regulatory body in facilities and activities or authorized parties beyond the interest necessary for regulatory purposes.
— Separation and effective independence of the regulatory body from the operating organization in the event that a department or agency of government is itself an authorized party that operates a regulated facility or facilities or conducts regulated activities.
— Emphasis on the independence of the regulatory body in orientation, training and responsibilities when new staff members are recruited from operating organizations. For example, the regulatory body could prohibit such staff from oversight of their previous employer for a defined period.

3.17. Paragraph 3.7 of Ref. [3] states:

"The State should ensure that the *regulatory body* and other *competent authorities* are adequately provided with the necessary authority, competence and financial and human resources to fulfil their assigned nuclear security responsibilities."

3.18. Such authority, competence and financial and human resources of the regulatory body and other competent authorities should include:

— Legal authority for establishment of regulations, authorization, inspection and enforcement with respect to the security of radioactive material;
— Sufficient personnel with the competence to effectively develop security regulations, assess operators' demonstration of compliance with security requirements, conduct security inspections and identify corrective actions, and recommend or take enforcement action as a result of non-compliance;
— Sufficient, regular, stable budgets to develop and maintain the foregoing competencies and staffing.

If personnel of the regulatory body responsible for protection and safety are assigned to perform security functions, they should receive appropriate training before assuming such responsibilities.

3.19. Paragraph 3.8 of Ref. [3] states:

"The State should establish requirements in accordance with national practices to ensure appropriate protection of specific or detailed information, which could compromise the security of *radioactive material, associated facilities* and *associated activities* if the information were disclosed."

3.20. States should designate the types of sensitive information[5] that are of security concern and should be protected. These types of information may include:

— Details of the security measures in place for radioactive material, associated facilities or associated activities, including information on guard and response forces;
— Information relating to the quantity, form and location of radioactive material, including radioactive material accounting information;
— Details of all computer based systems, including communication systems and instrumentation and control systems that process, handle, store and/or transmit information that is directly or indirectly important to safety or security;
— Response plans;
— Personal information about employees, vendors and contractors;

---

[5] As defined in Ref. [1], sensitive information is information, "in whatever form, including software, the unauthorized disclosure, modification, alteration, destruction, or denial of use of which could compromise nuclear security."

— Threat assessments and information;
— Details of vulnerabilities or weaknesses that relate to the above topics;
— Historical information on any of the above topics;
— Dates of future movements of radioactive material, especially between sites, including replacement of radioactive sources.

3.21. Information security refers to the system, programme or set of rules in place to ensure the confidentiality, integrity and availability of information in any form [18]. Reference [18] provides more comprehensive guidance on information security requirements, including on the establishment of a framework for securing sensitive information.

3.22. Access to sensitive information related to security of radioactive material, associated facilities and associated activities should be provided only to authorized individuals who have an operational need to know the information.

3.23. Individuals who possess sensitive information related to security of radioactive material, associated facilities and associated activities should be subject to regulatory requirements to protect the information from unauthorized disclosure and to report any actual or suspected unauthorized release, compromise or failure to protect sensitive information.

3.24. Paragraph 3.9 of Ref. [3] states:

> "The State should ensure that measures, consistent with national practices, are in place to ensure the trustworthiness of persons with authorized access to sensitive information or, as applicable, to *radioactive material, associated facilities* and *associated activities*."

3.25. The State should require the regulatory body to verify the trustworthiness of its staff who have access to sensitive information. In addition, the State should authorize and direct the regulatory body to require operators to establish policies and procedures to confirm through a background check the trustworthiness of individuals authorized for unescorted access to radioactive material or access to sensitive information. The regulatory body should ensure the availability of arrangements to enable operators to implement this requirement, such as referrals to law enforcement or other external agencies. In some States, this referral process may require facilitation by the regulatory body or another competent authority. The regulatory body or other competent authority should require that the results of trustworthiness verifications are appropriately protected as sensitive information.

3.26. States and regulatory bodies may need to establish laws or regulations to define minimum requirements, standards and scope for background checks and to establish penalties for misrepresenting material facts during background checks. States and regulatory bodies should also establish a framework that provides the capability to search criminal and counterterrorism databases as part of the background check. The details of these arrangements will vary depending upon the State's legislative and regulatory framework.

3.27. Paragraph 3.10 of Ref. [3] states (citation omitted):

"The State should establish, develop and maintain a national register of *radioactive material* over thresholds defined by the State. This national register should, as a minimum, include Category 1 and 2 radioactive sealed sources, as described in the Code of Conduct on the Safety and Security of Radioactive Sources. Other *radioactive material* could, as appropriate, be included in this register."

3.28. The State should authorize and require the regulatory body or other competent authority to establish, develop and maintain a national register of radioactive material. As recommended in Ref. [3] and described in Ref. [5], the register should, at a minimum, include all Category 1 and 2 radioactive sources, but could also include Category 3 radioactive sources or any other radioactive material which the State has determined is to be included in the national register. Such a register might already have been established for safety purposes. Information which may be included for each entry in the register includes the following, as applicable:

— Authorized person (licensee) and associated contact information;
— Radioisotope(s);
— Physical/chemical form;
— Weight/volume;
— Activity and date of measurement;
— Category/security level;
— Unique identifier of the radioactive source;
— Manufacturer source certificate;
— Location;
— Type of radioactive material (sealed source, unsealed material, etc.);
— Practice or use;
— Device in which the radioactive material is housed, including model number;
— Device serial number;
— Manufacturer of the device and associated contact information;

— Manufacturer and supplier of the radioactive material and associated contact information;
— Date of supply of radioactive material;
— Intended design lifetime of the radioactive material and/or device;
— Photograph of the device and/or radioactive material;
— Authorization (licence) number;
— Authorization (licence) termination date.

3.29. Each operator should be required to maintain an inventory which includes, at a minimum, all Category 1 and 2 radioactive sources. Annually, or more frequently, as specified by the regulatory body, the operator should be required to verify that the inventory is complete and accurate, and to adjust the inventory to reflect any discrepancies identified. The operator should be required to report these inventory results to the regulatory body or other competent authority, as applicable, for inclusion in the national registry of radioactive material. The operator should also be required to report receipts, transfers and disposition of radioactive material, either prospectively or within a specified period after the receipt or transfer occurs.

Regulatory body

3.30. Paragraph 3.11 of Ref. [3] states:

> "The *regulatory body* should implement the legislative and regulatory framework and authorize activities only when they comply with its nuclear security regulations. Where it is required, the security plan ... can be used by the regulatory body in its determination for issuance of an authorization."

3.31. The regulatory body should define the requirements for the security of radioactive material to be met prior to the authorization of activities involving this material and establish a process for review and approval (or denial) of applications for new authorizations and renewals of or amendments to existing authorizations. As previously noted, safety and security authorizations could be conducted in a single, integrated process or separately. An authorization that includes both safety and security could more readily address the safety–security interface.

3.32. Radioactive material above a certain activity threshold as defined by the State should be subject to authorization at all stages of its life cycle. All authorizations may be subject to amendment, renewal, revocation or suspension as determined to be necessary by the regulatory body, in conformance with established procedures and criteria. Each authorization should include:

— A reference to the legislation and regulations relevant to the activity or activities authorized;
— Specification of the activity or activities authorized;
— Any constraints regarding the activities, such as requirements, conditions, location or time limits.

3.33. The regulatory body's assessment of each application for an authorization should include a review of the security measures proposed by the applicant. If the regulatory body identifies any deficiencies, it should ensure that these deficiencies are corrected and that the final security measures are verified to be acceptable in accordance with established criteria and procedures.

3.34. When required by the regulatory body, based on a graded approach, the security plan should be one of the documents the applicant submits to the regulatory body as part of the authorization process. Compliance with the approved security plan should be a condition of the authorization once it is granted. The authorization itself should be an enforceable instrument, authorizing an activity or activities subject to compliance with authorization conditions and applicable legislation and regulations.

3.35. Paragraph 3.12 of Ref. [3] states:

"The regulatory body should verify continued compliance with nuclear security regulations and relevant authorization conditions, notably through periodic inspections and ensuring that corrective action is taken, when needed. Inspections of security measures implemented by an *operator* … could be performed together with inspections for verifying compliance with other regulatory requirements, such as radiation protection and safety. The security plan could be referred to by the *regulatory body* for these activities."

3.36. The regulatory body should develop and implement a programme of security inspections of facilities and activities to verify that the operator is in compliance with applicable regulatory requirements and with the conditions specified in the authorization. This programme should specify the types of regulatory inspection, including scheduled inspections and unannounced inspections. The frequency and depth of inspections should be commensurate with the security risks associated with the facility or activity, in accordance with a graded approach. Security could be addressed as part of radiation protection and safety inspections, provided that the inspectors are appropriately trained and qualified in security.

3.37. The regulatory body should record the results of inspections and should take appropriate follow-on action, including enforcement actions as necessary. Results of inspections should be used as feedback for the regulatory process and should be provided to the operator. Inspection results which contain sensitive information related to security should be handled as such. Provision should be made to allow access by regulatory inspectors to any facility or activity at any time, within the constraints of ensuring operational safety and security at all times and other constraints associated with the potential for harmful radiological consequences.

3.38. The regulatory body should establish and implement an enforcement policy within the legal framework for responding to non-compliance by operators with regulatory requirements or with any conditions specified in the authorization (including provisions of the security plan which should have been made mandatory through the authorization process). In the event that risks are identified, including risks unforeseen in the authorization process, the regulatory body should require operators to take corrective actions.

3.39. The response of the regulatory body to non-compliance with regulatory requirements or with any conditions specified in the authorization should be commensurate with the security significance of the non-compliance, in accordance with a graded approach.

3.40. Enforcement actions by the regulatory body based on established criteria could include recorded verbal notification, written notification, imposition of additional regulatory requirements and conditions, written warnings, penalties and, ultimately, modification, suspension or revocation of the authorization. Regulatory enforcement may also entail prosecution, particularly in cases where the operator does not cooperate satisfactorily in addressing the non-compliance.

3.41. At each significant step in the enforcement process, the regulatory body should identify and document the nature of the operator's non-compliance with regulatory requirements and the period of time allowed for correcting them, and should communicate this information in writing to the operator.

3.42. The operator should be held accountable for remedying non-compliance, for performing a thorough investigation in accordance with an agreed timetable and for taking all the measures that are necessary to prevent recurrence of the non-compliance.

3.43. The regulatory body should confirm that the operator has effectively implemented any necessary corrective actions.

Operator

3.44. Paragraph 3.13 of Ref. [3] states:

"The legislative and regulatory framework should require that the *operator* …:

— Comply with all applicable regulations and requirements established by the State and the *regulatory body*;
— Implement security measures that comply with requirements established by the State and the *regulatory body*;
— Establish quality management programmes that provide:
  • Assurance that the specified requirements relating to nuclear security are satisfied;
  • Assurance that the components of the *nuclear security system* are of a quality sufficient for their tasks;
  • Quality control mechanisms and procedures for reviewing and assessing the overall effectiveness of security measure;
— Report to the *regulatory body* and/or to any other *competent authority*, all *nuclear security events* involving *radioactive material, associated facilities* and *associated activities* according to national practices;
— Cooperate with and assist any relevant *competent authorities* in case of a *nuclear security event*."

3.45. The regulatory body should assign operators the primary responsibility for designing, implementing and maintaining security systems for radioactive material in accordance with regulatory requirements. While operators may be permitted, depending on the applicable regulatory requirements, to contract with a third party to carry out actions and tasks related to the security of radioactive material, the authorized operator should retain the primary responsibility for regulatory compliance and the effectiveness of said actions and tasks. In some cases, the regulatory body may establish requirements for activities assigned to a contractor. The regulatory body should also require operators to ensure that contractors are suitably trained and that these personnel meet those regulatory requirements that would apply if such personnel were directly employed by the operator, including requirements pertaining to trustworthiness. Operators should be further required to ensure that contractors have appropriate information security systems in place.

3.46. The regulatory body should require that operators conduct periodic evaluations of facilities to verify that they are in compliance with all applicable security requirements and to assess the effectiveness of their security systems to identify weaknesses that should be corrected, providing opportunities for continuous improvement. For example, these evaluations could take the form of a vulnerability assessment, a detailed explanation of which is provided in Appendix III. The evaluations should be performed using relevant threat information provided by the regulatory body.

3.47. The regulatory body should require that operators establish security management systems based on a risk informed graded approach and integrated with their overall management systems. The security management system should ensure that:

— The security system is reliably operated and maintained, functions as intended, is effective and meets regulatory requirements.
— Personnel, procedures and equipment function are effectively integrated as a system.
— Policies and procedures are established that identify security as being of high priority.
— Radioactive material is adequately identifiable, traceable and periodically verified to be present at its authorized location.
— Incidents affecting the security system are promptly identified and corrected in a manner commensurate with their importance, including but not limited to:
    • Confirmation that security measures, pertaining both to the security system and to security management, are and remain effective as long as radioactive material is present;
    • Notification to, cooperation with and assistance to the regulatory body and other competent authorities in case of nuclear security events, as required by legislation or regulation.
— The responsibilities for security held by individual personnel are clearly identified and personnel are suitably trained, qualified and determined to be trustworthy.
— Clear lines of authority for decisions on security are established.
— Organizational arrangements and lines of communications are established that result in an appropriate flow of information on security within the entire organization.
— Sensitive information is identified and protected according to national regulations.
— Radioactive material is protected in accordance with the security plan.

3.48. Paragraph 3.14 of Ref. [3] states: "States are encouraged to cooperate and consult, and to exchange information on nuclear security techniques and practices, either directly or through relevant international organizations."

3.49. Each State should consider whether, under what circumstances and to what extent it will cooperate with other States, including the appropriate sharing of information and knowledge derived from its national nuclear security regime, having regard to the sensitive nature of nuclear security information and the need to protect it and share it on the basis of the State's national legal framework.

3.50. While some facility specific sensitive information should not be shared, other useful information such as good practices may be shared in workshops, training programmes and conferences. Information can also be shared through the IAEA without attribution.

3.51. Paragraphs 3.15 and 3.16 of Ref. [3] state respectively:

"States concerned should, in accordance with their national law, provide cooperation and assistance to the maximum feasible extent in the location and recovery of *radioactive material* to any State that so requests."

"For the purpose of reporting *nuclear security events*, States should consider establishing suitable arrangements to enable them to participate in relevant regional and international databases and international activities in accordance with their national legislation. One example is the IAEA's [Incident and] Trafficking Database (ITDB). Consideration should also be given to other bilateral and multilateral support arrangements."

3.52. Provision of timely information to States and the IAEA regarding missing or stolen radioactive material is important to assist with its location and recovery. Notification provided to States and the IAEA regarding nuclear security events involving radioactive material could also assist in identifying and addressing potential threats associated with the material involved. Information may be provided on a voluntary basis to the IAEA Incident and Trafficking Database [19]. States may also choose to use other mechanisms established by the IAEA for notifying other States, sharing information and receiving financial or technical support in the event of a nuclear or radiological emergency triggered by a nuclear security event such as the unauthorized removal of radioactive material [20, 21]. In the case of unauthorized removal of radioactive material, the affected State may benefit particularly from assistance from neighbouring States in locating and recovering the missing radioactive material, if it might have entered or passed through those States. Detection of the material will be dependent on the system(s) for detection of nuclear and other radioactive material out of regulatory control in the State where the material is or through which it is passed. Further information on a State's system for detection of nuclear and other radioactive material outside of regulatory control is provided in Refs [4, 22].

3.53. State points of contact for nuclear security are especially important in the case of unauthorized removal or sabotage, to facilitate communicating essential information promptly and accurately to neighbouring States and other concerned parties. Such communication could occur either directly or through the IAEA. State points of contact for nuclear security may also be useful in communicating other important nuclear security information relevant to security of radioactive material, associated facilities and associated activities, such as information about new threats of common concern. These points of contact are most useful when established in advance of a nuclear security event.

IDENTIFICATION AND ASSESSMENT OF THREATS

3.54. Paragraph 3.17 of Ref. [3] states:

> "The State should assess its national *threat* for *radioactive material, associated facilities* and *associated activities*. The State should periodically review its national *threat*, and evaluate the implications of any changes in the *threat* for the design or update of its *nuclear security regime*."

3.55. The design and evaluation of security systems should take into account the current national threat assessment for radioactive material, associated facilities and associated activities, and the relevant design basis threat(s) (DBT) and/or representative threat statement(s) (RTS).

3.56. The process for assessing the national threat for radioactive material, associated facilities and associated activities and using this information is depicted in Fig. 1 and discussed in the following subsections.

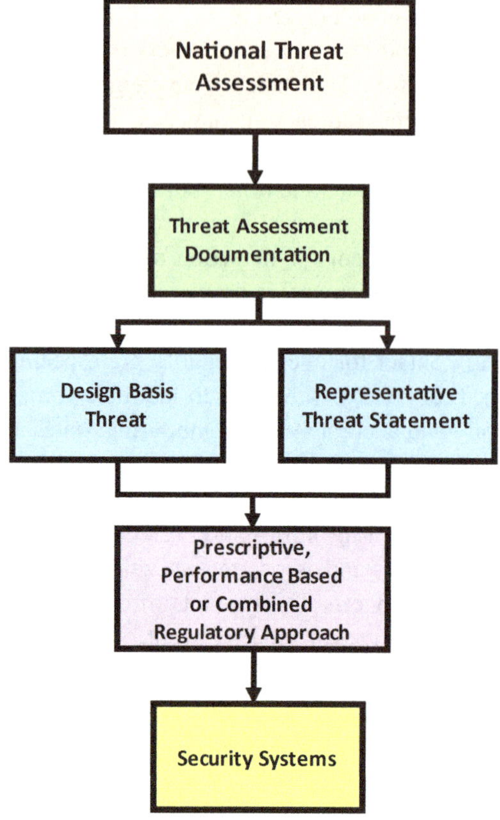

FIG. 1. Process for assessing and using threat information.

National threat assessment for radioactive material, associated facilities and associated activities

3.57. The national threat assessment for radioactive material, associated facilities and associated activities is an evaluation of the threats to radioactive material and associated facilities and associated activities — based on available intelligence, law enforcement and open source information — that describes the motivations, intentions and capabilities of potential adversaries to commit malicious acts. The national threat assessment for radioactive material, associated facilities and associated activities will be part of the national nuclear security threat assessment and may be part of a broader national threat assessment. For simplicity, the national threat assessment for radioactive material, associated facilities and associated activities is referred to as the 'national threat assessment' in the following text.

3.58. Sources of information for the national threat assessment should include, as appropriate, intelligence organizations, including security agencies, computer and information security organizations, law enforcement agencies, the International Criminal Police Organization–INTERPOL, the regulatory body for nuclear security and other competent authorities, customs and border agencies, the military services, shippers and carriers, official government reporting, incident reporting by operators, databases maintained by international organizations and other open sources. The national threat assessment should be updated on a regular basis or when circumstances make it necessary, such as when new information pertaining to threats is acquired.

3.59. Regulatory requirements for the design and evaluation of security systems should take into account the current national threat assessment to define the capabilities of the adversary, whether an insider or external, that the security system needs to address. Attributes and characteristics of adversaries that should be considered in the threat assessment are described in Ref. [23].

3.60. One method for using threat information in establishing regulatory requirements is for the competent authority responsible for the national threat assessment to provide an RTS, based on the results of the national threat assessment, to the regulatory body for its adaption and use in the development of its regulatory requirements for security of radioactive material, associated facilities and associated activities. Where this method is chosen, the regulatory body establishes regulations that require the operator to implement a security system which, based on the regulatory body's assessment, will protect against an adversary with the attributes and characteristics identified in the RTS.

3.61. Alternatively, the national threat assessment can be used to develop and apply a DBT, which the regulatory body could adapt and provide to the operator as a basis for the operator to design and implement a security system to meet regulatory requirements. Further guidance on national threat assessment and on defining a DBT based on a national threat assessment is given in Ref. [23].

3.62. In selecting whether to apply the national threat assessment through a DBT or an RTS, the State should consider several factors, including the severity of the consequences associated with malicious acts involving radioactive material in the State, the ability to establish effective security systems using each regulatory approach and the ability of the regulatory body to implement the different regulatory approaches, described in paras 3.84–3.86.

Design basis threat or representative threat statement

3.63. As described in more detail in Ref. [23], the analysis and decision making process involved in developing a DBT has three major phases:

— Screening the national threat assessment output for those threats with motivation, intention and/or capability to commit a malicious act involving radioactive material, associated facilities or associated activities;
— Collating the resulting screened list into a statement of representative attributes and characteristics of the postulated adversary;
— Tailoring the statement of representative threat attributes and characteristics on the basis of relevant policy considerations.

3.64. The output of this process is a detailed and comprehensive set of attributes and characteristics of threats against which operators are required to protect. The development of an RTS includes consideration of many of the same factors as those for a DBT, but in a less rigorous manner and perhaps involving fewer organizations. Nevertheless, a formal process for developing an alternative threat based protection should be undertaken, which should:

— Identify relevant threats from the national threat assessment;
— Assess the influence of policy factors;
— Document adversary capabilities in a threat statement that will be used by the regulatory body to define requirements for the design and evaluation of the security system.

3.65. If the State does not have sufficient resources to conduct a formal process of DBT development, or if the DBT process does not bring sufficient benefit in terms of reducing the risk associated with the radioactive material to be protected, then the State may choose to define an RTS.

3.66. A State may choose to define a DBT for high consequence radioactive material and an RTS for low consequence radioactive material.

3.67. The DBT or RTS and how they are developed will be specific to each State, owing to social, cultural and geopolitical differences. As with the national threat assessment, developing a DBT or RTS typically requires the combined efforts of domestic competent authorities such as intelligence and security agencies, law enforcement and the regulatory body and operators. The State should assign overall responsibility for preparing and maintaining the DBT or RTS to the regulatory body or another competent authority, as appropriate, depending on legislation and other national circumstances. The DBT or RTS should be reviewed at regular intervals and, when necessary, upon the availability of new information.

Output of DBT or RTS

3.68. According to Ref. [23] (footnote omitted):

> "The process of defining the DBT has two outcomes. The primary result is the DBT document. The DBT is that set of attributes and characteristics of threats for which the State organizations and the operators have protection responsibilities and accountability. However, the second result will identify those threats that are not appropriate for inclusion in a DBT but against which the State requires that protection should be reasonably ensured."

Such threats would be primarily countered by the State rather than by the operator, whose capabilities and/or resources for protection and response may be insufficient. The process of defining the RTS should have similar outcomes. As discussed in Section 5, the nature of the DBT or RTS information conveyed to operators will depend on the regulatory approach chosen.

Maintenance of the DBT or RTS

3.69. According to Ref. [23]:

> "A formal review process should be established to maintain the validity of a DBT. … The process should also include an assessment of quickly evolving threats that have to be dealt with urgently. In such circumstances, it may be necessary to take additional security measures before the DBT has been formally reviewed. The manner in which emerging threats are addressed will vary from State to State."

A similar process should be established to maintain the validity of an RTS, if this approach is selected.

3.70. The process for review of a DBT or RTS and the participants involved would be the same as for the original DBT or RTS, unless changes in law or government organization require alteration of these arrangements. The output of the review should be a determination as to whether the current DBT or RTS continues to suffice or a revised DBT or RTS is necessary. If a new DBT or RTS is issued, the regulatory body should assess its security regulations and their implementation by operators to determine whether amended regulatory requirements or modifications to operators' security system are necessary to counter the newly defined threat.

3.71. Situations may arise outside the regular review process in which adversaries are demonstrated or suspected to possess new or unexpected capabilities that are threatening enough to call for immediate action. The regulatory body and other competent authorities should put a process in place for the sharing of threat information among the competent authorities and with relevant operators. If an operator receives information on such a change in the threat through informal channels, the operator should inform the regulatory body and other competent authorities as appropriate. The credibility and relevance of the information and the severity of the potential impact of the change in the threat should be used to determine how, and how urgently, the State and/or the operator needs to respond.

Insider threats

3.72. The national threat assessment and the DBT or RTS, as applicable, should address the insider threat to radioactive material and associated facilities.

3.73. An insider is an individual with authorized access to associated facilities or associated activities or to sensitive information or sensitive information assets, who could commit or facilitate the commission of "criminal or intentional unauthorized acts involving or directed at *nuclear material, other radioactive material, associated facilities,* or *associated activities,* and other acts determined by the State to have an adverse impact on nuclear security" [1]. As described in more detail in Ref. [24], insider threats possess at least one of the following attributes that provide advantages over external threats when attempting malicious activities:

(a)   Authorized access: Insiders have authorized access to the areas, equipment and information needed to perform their work.
(b)   Authority: Insiders are authorized to conduct operations as part of their assigned duties and might also have the authority to direct other employees.
(c)   Knowledge: Insiders might possess knowledge of the facility or systems, ranging from limited to expert knowledge.

These attributes could also include access to or knowledge of sensitive information or sensitive information assets. Employees might also be susceptible to coercion, and operators should acknowledge this potential vulnerability.

3.74. The general methods described in Ref. [24] should be applied, using a graded approach, to protect against insider threats to radioactive material, associated facilities and associated activities. Insider threats can be addressed through technical measures, such as video surveillance and accounting, as well as security management measures, such as access control, trustworthiness verification and information protection. In addition, nuclear security culture plays a key role in ensuring that individuals, organizations and institutions remain vigilant and that sustained measures are taken to counter insider threats [25].

Increased threat

3.75. Security systems should be designed to be effective in countering any threat discerned by the national threat assessment, or any threat as determined by the DBT or RTS process. However, the regulatory body should also require that the operator make arrangements to ensure that security systems can be temporarily strengthened during times when a threat suddenly increases, including the introduction of additional security management measures. The operator should periodically test such measures and include them in the security plan.

3.76. To the extent that an increased threat is beyond that identified in the DBT or RTS, the primary responsibility for countering it is likely to rest with the State.

Evaluation methods

3.77. There are a number of methods for evaluating the effectiveness of a security system in protecting against identified threats. One such method is by means of a vulnerability assessment (VA). A VA can be specific or general in nature, and can be conducted by the operator to demonstrate system effectiveness (compliance) against the requirements specified in the State's regulatory framework or by the State's regulatory body to verify the operator's compliance. Vulnerability is assessed against the basic security functions of detection, delay and response, discussed further in Section 4, to ensure that the risks associated with malicious acts against radioactive material and associated facilities and activities, as defined by the State, are managed to an acceptable level. Additional information on how to conduct a VA can be found in Appendix III.

Risk based nuclear security systems and measures

3.78. Paragraphs 3.19 and 3.20 of Ref. [3] state respectively:

"The State should follow a structured risk management approach to reduce the risks of *malicious acts* to an acceptable level. The State should assess the potential *threats*, the potential consequences and the likelihood of *malicious acts*, and then develop a legislative and regulatory framework that provides for efficient and effective security measures to address the *threat*."

"The State should decide what level of risk is acceptable and what level of effort is justified to protect *radioactive material, associated facilities* and *associated activities* against the *threat* so as to reduce the risk to an acceptable level, given the availability of resources, the benefit of the protected asset to society, and other priorities. The required security measures may take advantage of other measures established for radiological safety purposes."

3.79. A structured risk management approach taken by the State is intended to reduce the risk associated with malicious acts to an acceptable level by evaluating the threat and potential harmful radiological consequences of such acts and ensuring that appropriate security measures are put into place.

3.80. The purpose of such an approach is to focus on reducing the likelihood of adversary success at completing malicious acts that could result in harmful radiological consequences. As described in more detail in Section 5, the regulatory body should establish three graded security levels for radioactive material, where each level is associated with a set of security requirements of increasing stringency. The regulatory body should assign radioactive material to a given security level based primarily on the potential harmful radiological consequences resulting from successful use of that material in a malicious act.

3.81. In addition to harmful radiological consequences, a malicious act could result in indirect consequences such as mass panic, psychological effects and loss of confidence in the industry using the radioactive material. While recognizing that all of these consequences are possible, this publication only takes into account the harmful radiological consequences of a malicious act when discussing the risk management process. However, States may consider these other indirect consequences when defining the acceptable level of risk within their territory.

3.82. Paragraph 3.21 of Ref. [3] states: "The *regulatory body* should establish regulations based on a prescriptive approach, a performance based approach or a combined approach in order to achieve the objectives of the nuclear security regime".

3.83. There are three possible approaches to establishing security regulations: a prescriptive approach, a performance based approach or a combined approach. The approach selected by the regulatory body should take into account its own capabilities and resources, the capabilities and resources of the operators that it regulates, the range of material that should be secured and the national legislative and regulatory framework.

3.84. In a prescriptive approach, the regulatory body establishes a set of specific security measures that it has determined provide an acceptable level of security against the threat as defined by the threat assessment and the DBT or RTS. A prescriptive approach has the advantage of simplicity in implementation, both for the regulatory body and for operators. The disadvantage of this approach is its relative lack of flexibility. For example, an operator's security system could be in compliance with prescriptive requirements but not fully address the actual vulnerabilities of the operator's radioactive material to particular threats.

3.85. In a performance based approach, the regulatory body defines security objectives on the basis of the threat assessment and the DBT or RTS, and requires that the operator design and implement a combination of security measures that can meet these objectives. The advantages of this approach are that it recognizes that an effective security system could be composed of a range of security measures, and that each operator's circumstances may be unique. The greater flexibility of the performance based approach also reduces the need for changes to regulations when new threats are identified. However, for this approach to be successful, both the operator and the regulatory body need to have sufficient personnel who possess high levels of security expertise.

3.86. In a combined approach, elements are drawn from both the prescriptive and performance based approaches. There are many possible versions of the combined approach. For example, the regulatory body could establish a set of security measures from which the operator could choose, and require the operator to demonstrate that the security system as a whole — developed by the operator using a subset of these security measures — meets the applicable security objectives defined by the regulatory body. Alternatively, the regulatory body could use a performance based approach for radioactive material with the greatest potential for harmful radiological consequences from malicious use and a prescriptive approach for material with less potential for harmful radiological consequences. The main advantage of the combined approach is the flexibility it allows the regulatory body in adjusting regulatory requirements to meet the specific needs and constraints of the operator. The three approaches are discussed in further detail in Section 6.

Use of alternative technologies or practices

3.87. Paragraph 3.22 of Ref. [3] states:

"The State should consider ways of reducing the nuclear security risk associated with *radioactive material*, particularly *radioactive sources*, for example by encouraging the use of an alternative radionuclide, chemical form, or non-radioactive technology, or by encouraging device designs that are more tamper resistant."

3.88. Consideration should be given to encouraging the use of newly developed or existing alternative technologies or operational practices in any application where the alternative technology or practice may reduce the security risk associated with this radioactive material. These technologies or practices, where technically and economically feasible, could rely, for example, on the use of:

— Another form of the same radionuclide, for example the use of caesium ceramics rather than caesium chlorides;
— An alternative radionuclide, for example the use of ^{3}H in radioluminescent devices in place of the traditionally used ^{226}Ra;
— A non-radioactive technology, for example the replacement of ^{137}Cs blood irradiators with X ray devices in some cases;
— Non-radiological techniques, for example the use of electronic gauges instead of level or density gauges containing ^{137}Cs or ^{60}Co sources;
— Modified operational practices, for example in industrial radiography moving items to be tested from the job site to a secure permanent facility.

3.89. Consideration should also be given to the use of more tamper resistant designs for devices using radioactive material, which can increase the time it takes to access and remove radioactive material from a device. The additional delay provided by these measures increases the time available for response forces to respond to an attempted or actual unauthorized removal of radioactive material. For example, the addition of hardware such as difficult to penetrate plates to better secure vulnerable maintenance locations may be considered. The use of specialty fasteners that require specialized tools for their installation and removal them may also be a valuable measure.

3.90. When investigating potential substitutes for radioactive material, both the advantages and disadvantages of the substitutes should be considered. While the throughput may be higher using some alternative technologies, the reliability of alternative technologies may not be sufficient in every situation; notably, X ray devices usually rely on a dedicated power supply that might not always be available.

3.91. States should exchange information regarding alternative technologies and practices. For example, if the security of a device design has been enhanced in one State, other States may also benefit from becoming aware of this enhancement.

Graded approach

3.92. Paragraph 3.23 of Ref. [3] states: "The *regulatory body* should develop requirements by using a *graded approach* applying the principles of risk management including a categorization of *radioactive material*."

3.93. Security based categorization refers to the process of categorizing radioactive material based on its activity and/or use, assigning an appropriate security level, and making adjustments to the security level and resulting security measures based on specific factors or considerations. The process is illustrated in Fig. 2 and described in more detail in Section 5.

3.94. For protection against unauthorized removal of radioactive material or sabotage, the State should consider the potential harmful radiological consequences of acts involving particular categories or security levels of radioactive material and apply a graded approach when developing regulatory requirements.

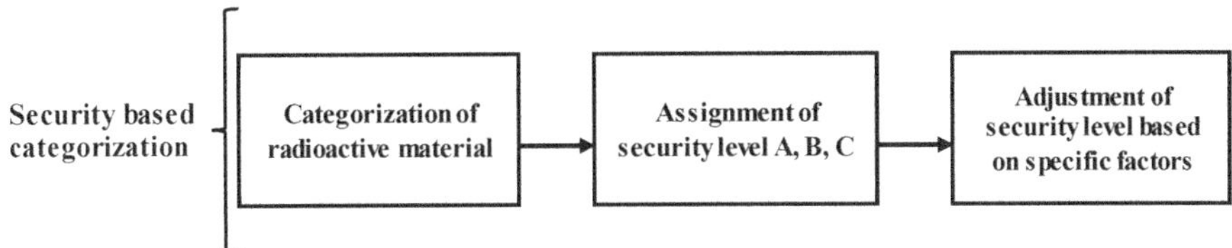

FIG 2. Security based categorization.

Defence in depth

3.95. Paragraph 3.24 of Ref. [3] states:

> "The *regulatory body* should develop requirements based on the concept of *defence in depth*. Security requirements for *radioactive material* require a designed mixture of hardware (security devices), procedures (access control, follow-up, etc.) and facility design."

3.96. The regulatory body should require that the defence in depth approach be used in the design of security systems for the nuclear security functions of detection, delay and response and in the implementation of security management. To the extent appropriate given the graded approach, the system design should include independent measures so that failure of one capability does not mean loss of a function. For example, both observation by personnel and electronic measures to detect intrusion into the facility can be used as detection measures. Delay measures may consist of multiple, independent and diverse physical barriers — such as fences, barricades, hardened buildings, hardened doors, cages and tie-downs — each of which must be overcome to gain access to the target. Response could be provided by both on-site guards and local police response. Security management measures should also incorporate the concept of defence in depth, where appropriate. For example, access control measures could include both a swipe card and a personal identification number.

3.97. By combining the principles of the graded approach and defence in depth when designing and implementing security measures for detection, delay and response, the operator may choose to use more layers and more effective components for higher consequence targets than for lower consequence targets.

INTERFACES WITH THE SAFETY SYSTEM

3.98. Paragraphs 3.25–3.28 of Ref. [3] state respectively:

> "Recognizing that both safety and security have a common aim — to protect persons, society and the environment from harmful effects of radiation — a well coordinated approach in safety and in security is mutually beneficial, the State should ensure that:
>
> — Consultation and coordination are maintained between those responsible for safety and security to ensure efficient security of

radioactive material and to ensure that regulatory requirements are consistent, especially when responsibility for safety and security is assigned to different *competent authorities*;

— Major decisions regarding safety and security require participation of experts in safety and in security on a continual basis;

— The safety and security interfaces should be strengthened by building safety culture and *nuclear security culture* into the management system."

"The State should ensure that a balance is maintained between safety and security throughout the *nuclear security regime*, from the development of the legislative framework to implementation of security measures."

"The *competent authorities* should ensure that security measures for *radioactive material, associated facilities* and *associated activities* take into account those measures established for safety and are developed so that they do not contradict each other, during both normal and emergency situations."

"The *competent authorities* working with the *operator* should ensure to the extent possible that security measures during a response to a *nuclear security event* do not adversely affect the safety of the personnel. Security personnel should manage their actions in a way that maintains the safety of all potentially affected persons, whether on or off-site."

3.99. The interfaces between safety and security should be taken into account at both the State and the operator level, as discussed in the following subsections.

State

3.100. The legislative and regulatory framework for the security of radioactive material, associated facilities and associated activities should take into account the existence, where applicable, of the legislative and regulatory framework for safety, including emergency preparedness and response, and radiation protection. Requirement 12 of Ref. [15] also addresses interfaces between safety and security.

3.101. Assigning the responsibility for both safety and security to a single regulatory body could help optimize resources and facilitate an integrated system of protection and control through authorization, inspection and enforcement processes. It could also simplify the regulatory body's ability to coordinate safety and security when developing regulations.

3.102.   There should be regular, systematic cooperation and information sharing between personnel in the regulatory body responsible for the development and implementation of safety requirements and those responsible for the development and implementation of security requirements. This cooperation and information sharing could include, but is not limited to:

— Consideration of both safety and security in the authorization processes for each area, including during categorization of radioactive material and inclusion of requirements for accounting and inventory;
— Review of proposed safety and security requirements to ensure their compatibility with one another;
— Shared inspections, as much as the protection of information allows for them;
— Joint assessment of the emergency plans and security plans provided by the operators to ensure compatibility and consistency;
— Involvement of safety specialists in the development of security requirements, and vice versa;
— Establishment of working groups dealing with specific technical interfaces.

3.103.   The working groups established to deal with specific technical interfaces may include, in addition to the regulatory bodies and as appropriate, technical support organizations, personnel from intelligence agencies; the ministries of interior, defence, transportation and foreign affairs; law enforcement, customs, coast guard and other agencies with security related responsibilities; and the ministries of health, environment or other agencies with responsibilities for safety, health or emergency preparedness and response. Working groups may also include senior management meetings to deal with major issues such as import–export controls and, as appropriate, ministerial arbitration in case of a remaining disagreement.

3.104.   Dedicated methods need to be developed to ensure the transparency of information pertaining to safety issues and to protect the information that is of a security concern. An integrated safety and security culture should be established within the regulatory body. Technical solutions should also be developed so that personnel with responsibilities related to the safety and security of radioactive material have access to the information needed to fulfil their duties, for example the data included in the national inventory of radioactive material.

Operators

3.105.   A good practice for establishing and maintaining an effective interface between safety and security is for the operator to implement safety and security measures in such a way that they are mutually supportive. For example, safety procedures used to prevent safety incidents could also support security. Safety and security measures should also be designed in such a manner that safety measures do not compromise security and security measures do not compromise safety.

3.106.   In many cases, the operator's staff dealing with safety issues will also deal with security issues. In those cases, the integration of safety and security is likely to be more easily achieved.

3.107.   When safety and security are not dealt with by the same staff within the operator's organization, safety and security specialists should be organized so that the interfaces between safety and security are well understood and managed. Senior management should participate in safety–security interface meetings and ensure that neither safety nor security is compromised by the other. Outcomes of safety–security interface meetings should be recorded. Security staff should have adequate knowledge of radiation protection requirements and related issues, and similarly, safety staff should be familiar with those security measures that are implemented in their work environment.

3.108.   Specific situations where safety–security interfaces should be addressed include:

— Maintenance of devices containing radioactive sources.
— Replacement of radioactive sources.
— Conduct of inventories of radioactive sources (or, when required, radioactive materials).
— Any change in the safety or security systems or in the design/characteristics of the facility (location of radioactive material, type of devices, access control, etc.). Such modifications should always be analysed from both a safety and a security point of view before being implemented. Where potential adverse impacts are identified, the operator should communicate them to appropriate personnel within the organization and consider alternative measures or take compensatory and/or mitigating actions.
— Access control (including the definition of access control areas) and access to information.
— Consideration of the radiation protection programme in the development of the security plan.

3.109. The operator should recognize safety–security interface issues and manage them appropriately during normal operations as well as during emergencies. Emergencies, whether due to a safety event or a nuclear security event, are of particular concern. Management of these interfaces should include:

— Ensuring, as far as possible, that arrangements for emergency management have been taken into account in the development of the security system;
— Coordinating and integrating security plans with emergency plans;
— Developing and conducting regular shared exercises between safety and security to test the coordinated plans and arrangements;
— Ensuring, as far as possible, that security response forces have adequate knowledge of radiation protection policies and procedures, and that they are designated, depending on their duties, as emergency workers and appropriately protected as described in Refs [9, 10, 26];
— Maintaining security to the extent possible during an emergency.

SUSTAINING THE NUCLEAR SECURITY REGIME

3.110. Paragraphs 3.29–3.32 of Ref. [3] state respectively:

"The State should commit the necessary resources, including human and financial resources, to ensure that its *nuclear security regime* is sustained and effective in the long term to provide adequate nuclear security for *radioactive material*."

"The State should promote a *nuclear security culture*."

"All organizations and individuals involved in implementing nuclear security should give due priority to the *nuclear security culture* with regard to *radioactive material*, to its development and maintenance necessary to ensure its effective implementation in the entire organization."

"The foundation of a *nuclear security culture* should be the recognition that a credible *threat* exists, that preserving nuclear security is important, and that the role of the individual is important."

3.111. Sustainability is the set of principles and implementing actions incorporated into the nuclear security regime that support its continuing effectiveness against a defined threat at both the national and the operational levels.[6] More detailed guidance on the key principles and actions for sustaining a nuclear security regime is provided in Ref. [27].

3.112.   Operators should promote a strong and effective security culture at all levels of operator staff and management of facilities with radioactive material. Reference [28] provides more detailed guidance on nuclear security culture.

PLANNING AND PREPAREDNESS FOR AND RESPONSE TO NUCLEAR SECURITY EVENTS

3.113.   Paragraph 3.33 of Ref. [3] states: "The *regulatory body* should ensure that the *operator*'s security plan includes measures to effectively respond to a *malicious act* consistent with the *threat*."

3.114.   The regulatory body should require the operator to include measures in its security plan that ensure a timely and effective response to a suspected, attempted or actual malicious act involving radioactive material within the facility.

3.115.   The regulatory body should ensure that facility specific response measures in the operator's security plan are consistent with those developed at the State and local levels. Any nuclear security event at the facility with off-site consequences should be managed in a coordinated and integrated manner that takes into account all organizations involved in the response, including the State, regulatory body, operator and other local/national response authorities.

3.116.   Arrangements should be made to ensure, as far as practicable, the continued effectiveness of the security system during the response to a nuclear security event, including through coordinated and integrated planning in the development and exercise of appropriate response measures by the State, regulatory body, operator and other local/national response authorities.

3.117.   Response measures should be developed based on information contained in the threat assessment and taking into consideration all foreseeable scenarios. These measures should be periodically exercised, reviewed and revised as necessary. The regulatory body should require the operator to implement appropriate response measures, for example by including the implementation of such response measures in the authorization conditions.

---

[6] The operational level includes those nuclear security systems implemented at a facility or in connection with any other activity where radioactive material is present.

3.118. The operator's security plan should take into account emergency arrangements established to effectively respond to a nuclear or radiological emergency in line with Refs [9, 10, 26] and based on a graded approach.

IMPORT AND EXPORT OF RADIOACTIVE MATERIAL

3.119. Paragraph 3.34 of Ref. [3] states:

"The State should take appropriate steps, including coordination between importer and exporter States prior to the transfer, to reduce the likelihood of *malicious acts* in connection with the import or export of quantities of *radioactive material* above thresholds that it defines. At a minimum, these steps should encompass requirements concerning Category 1 and 2 sealed *radioactive sources*, consistent with the Guidance on the Import and Export of Radioactive Sources".

3.120. Effective import–export control measures for radioactive material, and specifically Category 1 and 2 radioactive sources, serve several important security related purposes:

— Increasing importing State awareness of corresponding safety and security risks;
— Protecting against sources falling out of regulatory control during the export–import process and thus risking becoming lost, abandoned or stolen;
— Providing assurance that exported sources will be safely and securely managed throughout their life cycle.

3.121. Consistent with the Code of Conduct on the Safety and Security of Radioactive Sources [5] and the Guidance on the Import and Export of Radioactive Sources [8], the State should authorize and require the regulatory body or other competent authority to establish and implement a system for controlling the import and export of all Category 1 and 2 radioactive sources. States may consider extension of such measures to exports or imports of other radioactive material on a risk informed basis.

3.122. This system should include, as applicable:

— Nomination of a point of contact for facilitating communication between importing and exporting States related to import–export control of radioactive material.

— Establishment and implementation of procedures for the authorization and control of imports and exports that allow such imports and exports only if:

  • The recipient is authorized by the importing State to receive and possess the radioactive material;
  • The importing State has the capability to safely and securely manage the radioactive material;
  • The exporting State has sought and received the importing State's consent to the import (for Category 1 radioactive sources only);
  • The exporting State has notified the importing State prior to shipment.

— Provisions for consideration of authorizing imports and exports if one or more of the foregoing provisions cannot be followed owing to exceptional circumstances.

— Submission of the State's responses to the Importing and Exporting State Questionnaire associated with the Guidance on the Import and Export of Radioactive Sources [8], as well as any updates of those responses to the IAEA through official channels.

DETECTION OF NUCLEAR SECURITY EVENTS

3.123.   Paragraph 3.35 of Ref. [3] states:

"The *regulatory body* should establish requirements for *operators* … to have appropriate and effective security measures to detect *nuclear security events* and to report any such event promptly with the aim of providing a timely response. These requirements should consider those made in IAEA Nuclear Security Series No. 15, Nuclear Security Recommendations on Nuclear and Other Radioactive Material out of Regulatory Control".

3.124.   The regulatory body should require the operator to establish, test and implement measures to detect and respond to nuclear security events, using a graded approach and in cooperation with State and local level emergency and response plans. These measures should be documented in the operator's security plan or in a stand-alone response plan.

3.125.   The regulatory body should also establish requirements addressing when and how the operator is to report nuclear security events, including those addressing procedures for:

— Determining whether the event detected is a nuclear security event;

— Timely reporting to the regulatory body, the competent authority with responsibilities for radiological emergency response and law enforcement, as appropriate;
— Taking appropriate action to remedy or mitigate the circumstances;
— Investigating the event and its causes, circumstances, and actual and potential consequences, to prevent a recurrence of similar situations;
— Providing the regulatory body with a report within a specified period on the causes of the event, its circumstances and consequences, and on the corrective or preventive actions taken or to be taken.

The regulatory body should also require the operator to coordinate with the relevant competent authorities if the operator's radioactive material is lost, stolen or missing.

4. GUIDANCE ON THE SECURITY OF RADIOACTIVE MATERIAL

4.1. This section introduces and explains deterrence, detection, delay, response and security management, and provides guidance on integrating them into a security system, based on the recommendations contained in section 4 of Ref. [4]. Additional detailed guidance is provided in Sections 5 and 6 on establishing and implementing a regulatory programme for the security of radioactive material in use and storage and of associated facilities and associated activities.

SECURITY FUNCTIONS AND MEASURES

4.2. Security measures to address deterrence, the three security functions of detection, delay and response, and security management are addressed in detail in the sections that follow.

Deterrence

4.3. Deterrence is achieved if an adversary, otherwise motivated to perform a malicious act, is dissuaded from undertaking the attempt, for example because he or she estimates that the probability of success is too low or the potential negative consequences are too high.

4.4. Deterrence measures could include making the adversary aware of the presence of security measures in order to deter them from attempting a malicious act. However, communicating certain details of security measures might enable adversaries to circumvent or defeat the security system. Regulatory bodies and operators should consider how to balance this possibility with the potential deterrence value of alerting the adversary to the presence of security measures.

Detection

4.5. Detection is a process that begins with sensing a potentially malicious or otherwise unauthorized act (i.e. an alarm) and that is completed with an assessment of the cause of the alarm.

4.6. Detection and assessment can be achieved using different types of measures. For example, sensing unauthorized access can be accomplished through electronic sensors or visual observation. Sensing unauthorized removal can be accomplished through such means as tamper detection devices or visual observation or, after the fact, by accountancy records. Assessment can be performed through such means as remote video monitoring or visual observation.

Delay

4.7. Delay measures seek to slow down an adversary's attempt to complete a malicious act. Delay is considered to be the length of time, after detection, that an adversary needs to remove or sabotage radioactive material. For example, delay measures would slow down an attempt to gain unauthorized access to a location where radioactive material is present, or to remove or sabotage radioactive material, thereby providing more time for an effective response. Delay is typically added through the use of barriers or other physical obstacles that must be penetrated or defeated.

Response

4.8. Response encompasses the actions undertaken following detection of a nuclear security event to prevent an adversary from successfully completing an act of unauthorized removal or sabotage. Response activities, which could be performed by on-site guards or by off-site law enforcement, security or military personnel, seek to interrupt and defeat an adversary while the attempted unauthorized removal or sabotage is in progress in order to prevent its completion. When off-site responders are involved, the operator should coordinate with them in advance.

Security management

4.9. Security management addresses the establishment and implementation of policies, plans and procedures for the security of radioactive material, associated facilities and associated activities as well as the deployment of the necessary resources. Security management includes measures for access control, trustworthiness verification, information protection, preparation of a security plan, training and qualification of personnel, accounting, inventory and event reporting.

GUIDANCE ON THE SECURITY OF RADIOACTIVE MATERIAL IN USE AND STORAGE

Security system

4.10. A security system is an integrated set of nuclear security measures intended to prevent the completion of a malicious act. A malicious act consists of a sequence of actions undertaken by an adversary to obtain access to radioactive material for either sabotage or unauthorized removal.

4.11. The operator should design the security system to deter adversaries from attempting a malicious act and to prevent them from completing such a malicious act through the implementation of detection, delay and response measures. The security system should also include security management measures for the integration of people, procedures and equipment through the application of administrative measures.

4.12. *Detection before delay.* The operator should implement security measures so that an adversary would encounter detection measures prior to encountering delay measures. The intent of delay measures is to provide response personnel with sufficient time to deploy and interrupt the adversary's efforts to complete a malicious act. If an adversary is given the opportunity to overcome barriers and other obstacles intended to delay his or her progress towards the target prior to encountering intrusion sensors or other means of detection, the adversary will have completed some of the necessary tasks before being detected. This could increase the adversary's chances of succeeding in removing or sabotaging the radioactive material before the arrival of response personnel.

4.13. *Detection needs to be assessed.* Most means of detection provide an indirect indication of a potential malicious act. Therefore, when an alarm or other indirect indication that a malicious act might be under way is triggered, an assessment should be undertaken to determine its cause. There is always some uncertainty as to the cause of alarms. Alarm assessment requires human observation and judgement, through deployment of response personnel to investigate the cause of the alarm or through use of remote video systems. To prevent adversaries from exploiting any delay between detection and assessment, alarms should be assessed as soon as possible.

4.14. *Delay should be greater than assessment time plus response time.* Assessment activities should be undertaken soon enough after detection and performed quickly enough to enable response personnel to interrupt the adversary prior to completion of the unauthorized removal or sabotage. Thus, the time to assess and respond needs to be shorter than the time required for the adversary to defeat subsequent delay measures. This relationship of the functions of detection, delay and response is known as timely detection.

4.15. *Response should be adequate.* There should also be a sufficient number of response personnel with the tactics, skills and training necessary to defeat adversaries possessing the capabilities identified in relevant threat information.

4.16. *Balanced protection is needed.* The security system should be designed to provide adequate protection against all defined threats along all possible adversary pathways to the target. Along any possible pathway, detection measures, delay times and resulting responses should combine to protect the target.

4.17. *Defence in depth is needed.* A security system should employ the principle of defence in depth, such that several layers and methods of protection (structural, technical, personnel and organizational) need to be overcome or circumvented by an adversary in order to achieve his or her objective.

GUIDANCE ON THE SECURITY OF RADIOACTIVE MATERIAL IN TRANSPORT

4.18. Details on the development of these security requirements as well as the design of a security system for the transport of radioactive material are provided in Ref. [11], which explicitly addresses the recommendations contained in Ref. [3], paras 4.26–4.38.

5. ESTABLISHING A REGULATORY PROGRAMME FOR THE SECURITY OF RADIOACTIVE MATERIAL

5.1. This section provides guidance to regulatory bodies on how to develop or enhance regulatory programmes to address the security of radioactive material.

5.2. Many States already have a regulatory programme in place that covers activities such as authorization, inspection and enforcement for the safety of radioactive material. The purpose of a regulatory programme for the security of radioactive material is to bring the security risk to a level judged acceptable by the State. Safety and security measures should be designed and implemented in an integrated manner so that they strengthen one another to the extent possible and do not compromise one another.

5.3. The method for establishing a regulatory programme for the security of radioactive material described in this section involves three steps for the regulatory body:

— *Step 1:* Establish graded security levels, with corresponding goals and sub-goals for each security level.
— *Step 2:* Determine the security level applicable to a given radioactive material.
— *Step 3:* Establish regulatory requirements, using a prescriptive, performance based or combined approach.

5.4. The following sections provide more specific guidance on the means of implementing each step. However, States and their regulatory bodies could implement these steps or establish a regulatory programme in a different manner, as judged necessary to address national practice and circumstances.

STEP 1: ESTABLISH GRADED SECURITY LEVELS WITH CORRESPONDING GOALS AND SUB-GOALS

5.5. Applying a risk management approach to a nuclear security system means that the required degree of security system effectiveness is based primarily on the harmful radiological consequences that could result from a successful malicious act involving the particular radioactive material to be protected.

5.6. Three security levels (A, B and C) could be used to specify security system performance in a graded manner. Security level A would call for the highest degree of security system effectiveness while security levels B and C would call for progressively less stringent degrees of protection. This system is described in the remainder of this section.

5.7. If this approach is applied, the security system performance required by the regulator for each security level should be expressed as a goal. Such goals define the overall result that a security system needs to be capable of providing for radioactive material at that security level. The following goals apply to security levels A, B and C:

— *Security level A:* Provide a *high* level of protection of radioactive material against unauthorized removal.
— *Security level B:* Provide an *intermediate* level of protection of radioactive material against unauthorized removal.
— *Security level C:* Provide a *baseline* level of protection of radioactive material against unauthorized removal.

5.8. In order to meet the applicable goals for each security level, the security system needs to display an adequate level of performance for each of the security functions of detection, delay and response as well as for security management.[7] This level of performance can be expressed as a set of sub-goals associated with the performance of each of the functions and for security management. These sub-goals should state the outcome required by the regulatory body to result from the combination of measures applied for that function.

5.9. Malicious acts might involve either unauthorized removal of radioactive material or sabotage. While the security goals and recommended measures described in this section focus only on unauthorized removal of radioactive material, a security system whose performance achieves the applicable goals described here will also provide some capability to counter sabotage. If an approach to security of radioactive material such as the one described in this section is applied and the regulatory body becomes aware of a specific threat of sabotage against particular facilities, "the *regulatory body* should require additional or more stringent security measures to increase the level of protection against *sabotage*" [3]. The regulatory body may also choose to establish security levels, security goals and requirements that take into account the potential radiological consequences of sabotage.

---

[7] Deterrence achieved by a security system is difficult to measure. Consequently, it has not been assigned a set of goals and measures in this publication.

5.10. Security levels and goals and sub-goals associated with them are summarized in Table 3. Where a sub-goal is shown in Table 3 as the same for two or more security levels, it is intended that the sub-goal be met in a more rigorous manner for the higher security level.

TABLE 3. SECURITY LEVELS AND ASSOCIATED SECURITY GOALS AND SUB-GOALS BY SECURITY FUNCTION

Security function	Security level A	Security level B	Security level C
	Goal		
	Provide a *high* level of protection of radioactive material against unauthorized removal [a]	Provide an *intermediate* level of protection of radioactive material against unauthorized removal [a]	Provide a *baseline* level of protection of radioactive material against unauthorized removal [a]
	Sub-goals		
	Provide immediate detection of any unauthorized access to locations where radioactive material is present		
Detection	Provide immediate detection of any attempted unauthorized removal of radioactive material, including by an insider	Provide detection of any attempted unauthorized removal of radioactive material	Provide detection of unauthorized removal of radioactive material
	Provide immediate assessment of detection		
	Provide a means to detect loss of radioactive material through verification		
Delay	Furnish sufficient delay to provide a high level of protection against unauthorized removal of radioactive material	Furnish sufficient delay to provide an intermediate level of protection against unauthorized removal of radioactive material	Furnish sufficient delay to provide a baseline level of protection against unauthorized removal of radioactive material

TABLE 3. SECURITY LEVELS AND ASSOCIATED SECURITY GOALS AND SUB-GOALS BY SECURITY FUNCTION (cont.)

Security function	Security level A	Security level B	Security level C
	Goal		
	Provide a *high* level of protection of radioactive material against unauthorized removal [a]	Provide an *intermediate* level of protection of radioactive material against unauthorized removal [a]	Provide a *baseline* level of protection of radioactive material against unauthorized removal [a]
	Sub-goals		
Response	Provide immediate communication to response personnel		Provide prompt communication to response personnel
	Provide for immediate response with sufficient resources to interrupt and prevent the unauthorized removal of radioactive material	Provide immediate initiation of response to interrupt unauthorized removal of radioactive material	Implement appropriate action in the event of unauthorized removal of radioactive material
Security management	Establish a process for unescorted access to radioactive material and/or access to sensitive information Ensure trustworthiness and reliability of authorized individuals Provide access controls that effectively restrict access to radioactive material to authorized persons only Identify and protect sensitive information Provide a security plan Ensure training and qualification of individuals with security responsibilities Conduct accounting and inventory of radioactive material Conduct evaluation for compliance and effectiveness of the security system, including performance testing Establish a capability to manage and report nuclear security events		

[a] Achievement of these goals will also reduce the likelihood of a successful act of sabotage.

STEP 2: DETERMINE THE SECURITY LEVEL APPLICABLE TO RADIOACTIVE MATERIAL: SECURITY BASED CATEGORIZATION

5.11. If security levels are used to specify security system performance in a graded manner, the process of specifying an appropriate security level for radioactive material should consist of the following steps:

— Categorizing radioactive material based on its potential to cause harmful radiological consequences if used in a malicious act (including aggregation of radioactive material in a given location, as appropriate) (see paras 5.13–5.29);
— Assigning an appropriate security level to each category (see paras 5.30–5.32);
— Adjusting the security level based on specific factors or considerations (see paras 5.33–5.54).

5.12. The approach described here should be applied to all radioactive material, including radioactive sources, unsealed radioactive material and radioactive waste; however, it should be recognized that this approach is designed for radioactive sources, and should therefore be adapted where possible and as appropriate to suit the particular circumstances.

Categorization

Categorization of radioactive sources

5.13. IAEA Safety Standard Series No. RS-G-1.9, Categorization of Radioactive Sources [29], recommends a categorization system[8] based on a set of D values corresponding to "that quantity of radioactive material, which, if uncontrolled, could result in the death of an exposed individual or a permanent injury that decreases that person's quality of life" [30]. The D values for radioactive material were developed to establish requirements for an adequate level of preparedness for and response to a nuclear or radiological emergency, but take into account a number of defined exposure scenarios. These include those security scenarios resulting from the malicious use of radioactive material such as use in a radiological dispersal device, placement of an unshielded source in a public area and placement of radioactive material in a food or water supply.

---

[8] The categorization system described here is concerned with radioactive sources. This method could also be adapted for application to other radioactive material, depending on national considerations.

5.14. Two different D values are calculated for two different types of scenario. For those scenarios in which the radioactive material is not dispersed, a D_1 value for radionuclides is calculated, which is the activity "of a radionuclide in a source that if uncontrolled, but not dispersed... might result in an emergency that could reasonably be expected to cause severe deterministic health effects" [30]. For those scenarios in which the radioactive material is dispersed, a D_2 value is calculated, denoting the activity "of a radionuclide in a source that if uncontrolled and dispersed might result in an emergency that could reasonably be expected to cause severe deterministic health effects" [30]. The D value of a radionuclide is the lowest value of the D_1 and D_2 values for that radionuclide.[9]

5.15. In the system described in Ref. [29], radioactive material is categorized by taking the activity of the radioactive material (in TBq), A, and dividing it by the D value for the relevant radionuclide. The A/D value is referred to as the activity ratio. The category of radioactive material is then assigned to a category between 1 and 5 based on the value of the activity ratio, where 1 represents the highest level of danger and 5 the lowest. Radioactive sources in Category 1 can pose a very high risk to human health if not managed safely and securely. An exposure of only a few minutes to an unshielded Category 1 source might be fatal. Radioactive sources in Category 5 are the least dangerous; however, these sources should be kept under appropriate regulatory control. Categories and potential associated activity ratios are shown in Table 4 [29].

5.16. For example, a blood irradiator containing a ^{137}Cs source has an activity of 260 TBq. The D value of ^{137}Cs is 0.1 TBq. Therefore, taking the A/D ratio, 260 TBq/0.1 TBq = 2600. This means the A/D ratio $\geq$ 1000, so the radioactive source would be assigned to Category 1.

5.17. A State may choose to employ a different approach (e.g. consider different exposure pathways and/or dose rates) than the one used to calculate the D values described in Ref. [30] in order to categorize radioactive material for the purpose of assigning a security level.

5.18. While it may be appropriate to categorize radioactive material on the basis of its A/D ratio, it may also be convenient to assign a category on the basis of the intended application of the radioactive material [29]. Table 4 provides examples of categorization of radioactive material based on its application. For example,

---

[9] D values are also used as the basis for RS-G-1.9 [29]. The D values used in RS-G-1.9 [29] are the more stringent of the D_1 and D_2 values calculated in Ref. [30].

a blood irradiator may be assigned to Category 2 based on the A/D ratio, but assigned to Category 1 based on practice.

TABLE 4. CATEGORIES OF RADIOACTIVE SOURCES FOR COMMON APPLICATIONS

Category	Activity ratio (A/D) [a]	Applications [b]
1	$A/D \geq 1000$	Radioisotope thermoelectric generators Irradiators Teletherapy Fixed multibeam teletherapy (gamma knife)
2	$1000 > A/D \geq 10$	Industrial gamma radiography High/medium dose rate brachytherapy
3	$10 > A/D \geq 1$	Fixed industrial gauges that incorporate high activity sources [c] Well logging gauges
4	$1 > A/D \geq 0.01$	Low dose rate brachytherapy (except eye plaques and permanent implants) Industrial gauges that do not incorporate high activity sources Bone densitometers containing a radioactive source Static eliminators
5	$0.01 > A/D$ and $A >$ exempt [d]	Low dose rate brachytherapy eye plaques and permanent implant sources X ray fluorescence devices containing a radioactive source Electron capture devices Mossbauer spectrometry Positron emission tomography check sources

[a] This column can be used to determine the category of radioactive material purely on the basis of A/D. This might be appropriate, for example, if the facilities and activities are not known or are not listed, if radioactive material has a short half-life and/or is unsealed, or if radioactive material is aggregated (see RS-G-1.9 [29], para. 3.5).

[b] Factors other than A/D alone have been taken into consideration in assigning these applications to a category (see RS-G-1.9 [29], annex I).

[c] Examples are given in RS-G-1.9 [29], annex I.

[d] Exempt quantities are given in schedule I of GSR Part 3 [16].

5.19. The Code of Conduct on the Safety and Security of Radioactive Sources [5] (the Code) applies to radioactive sources that might pose a significant risk to individuals, society and the environment (i.e. Category 1–3 radioactive sources). The set of radionuclides included in the Code was developed based on national experiences and widespread uses of radioactive material at the time of the Code's publication in 2004; it is recommended that, at a minimum, this guidance should be applied to those radionuclides. These radioactive sources, their D values and activity thresholds for Categories 1–3 are listed in Table 5, which appears in annex I of the Code. The list of radionuclides provided in Table 5 should not be viewed as static, and could be modified to reflect fluctuations in industry and new needs which might evolve. For radionuclides not found in this table, the recommended D values are found in Ref. [30]. The regulatory body may choose to assign a category to these radionuclides based on the A/D ratio.

Categorization of other radioactive material

5.20. The categorization system described here could also be applied to radioactive waste, depending on the national context. The assignment of security levels to radioactive waste, if applicable, should follow the steps described in this guidance, with consideration of the classification system for radioactive waste described in IAEA Safety Standards Series No. GSG-1, Classification of Radioactive Waste [31]. The security levels, goals and sub-goals for radioactive waste would be as described in Table 3.

5.21. The categorization system could also be applied to nuclear material, depending on national circumstances. The D values for nuclear material are given in appendix III of Ref. [10] and the same process of categorization using the A/D ratio should be applied in order to categorize them based on the harmful radiological consequences of unauthorized removal for potential off-site exposure or dispersal. There are radionuclides for which D values are given as 'UL' or 'unlimited quantity'. This designation is for those radionuclides having very long half-lives and, therefore, low specific activities; very low energy radiation emissions; or a combination of both. It would therefore be impractical to use such material in a malicious act, and so its categorization for security purposes is unnecessary. For example, a pressurized water reactor fuel assembly might weigh as much as 660 kg. If this entire mass is assumed to be uranium (some percentage will be fuel cladding), the isotopic breakdown and A/D values for typical reactor fuel (4% enrichment in ^{235}U) are as shown in Table 6.

5.22. In this example, the A/D value is nominal for ^{234}U and ^{238}U and is significant for ^{235}U. However, the footnote to table 1 in Ref. [30] indicates that the D values

TABLE 5.   ACTIVITIES CORRESPONDING TO THRESHOLDS OF CATEGORIES

Radionuclide	Category 1 1000 × D		Category 2 10 × D		Category 3 D	
	(TBq)	(Ci)[a]	(TBq)	(Ci)[a]	(TBq)	(Ci)[a]
^{241}Am	6.E + 01	2.E + 03	6.E–01	2.E + 01	6.E – 02	2.E + 00
^{241}Am/Be	6.E + 01	2.E + 03	6.E – 01	2.E + 01	6.E – 02	2.E + 00
^{252}Cf	2.E + 01	5.E + 02	2.E – 01	5.E – 00	2.E – 02	5.E – 01
^{244}Cm	5.E + 01	1.E + 03	5.E – 01	1.E + 01	5.E – 02	1.E + 00
^{60}Co	3.E + 01	8.E + 02	3.E – 01	8.E + 00	3.E – 02	8.E – 01
^{137}Cs	1.E + 02	3.E + 03	1.E + 00	3.E + 01	1.E – 01	3.E + 00
^{153}Gd	1.E + 03	3.E + 04	1.E + 01	3.E + 02	1.E + 00	3.E + 01
^{192}Ir	8.E + 01	2.E + 03	8.E – 01	2.E + 01	8.E – 02	2.E + 00
^{147}Pm	4.E + 04	1.E + 06	4.E + 02	1.E + 04	4.E + 01	1.E + 03
^{238}Pu	6.E + 01	2.E + 03	6.E – 01	2.E + 01	6.E – 02	2.E + 00
^{239}Pu/Be [b]	6.E + 01	2.E + 03	6.E – 01	2.E + 01	6.E – 02	2.E + 00
^{226}Ra	4.E + 01	1.E + 03	4.E – 01	1.E + 01	4.E – 02	1.E + 00
^{75}Se	2.E + 02	5.E + 03	2.E + 00	5.E + 01	2.E – 01	5.E + 00
^{90}Sr (^{90}Y)	1.E + 03	3.E + 04	1.E + 01	3.E + 02	1.E + 00	3.E + 01
^{170}Tm	2.E + 04	5.E + 05	2.E + 02	5.E + 03	2.E + 01	5.E + 02
^{169}Yb	3.E + 02	8.E + 03	3.E + 00	8.E + 01	3.E – 01	8.E + 00
^{198}Au[*]	2.E + 02	5.E + 03	2.E + 00	5.E + 01	2.E – 01	5.E + 00
^{109}Cd[*]	2.E + 04	5.E + 05	2.E + 02	5.E + 03	2.E + 01	5.E + 02

TABLE 5. ACTIVITIES CORRESPONDING TO THRESHOLDS OF CATEGORIES (cont.)

Radionuclide	Category 1 1000 × D		Category 2 10 × D		Category 3 D	
	(TBq)	(Ci)[a]	(TBq)	(Ci)[a]	(TBq)	(Ci)[a]
^{57}Co[*]	7.E + 02	2.E + 04	7.E + 00	2.E + 02	7.E − 01	2.E + 01
^{55}Fe[*]	8.E + 05	2.E + 07	8.E + 03	2.E + 05	8.E + 02	2.E + 04
^{68}Ge[*]	7.E + 02	2.E + 04	7.E + 00	2.E + 02	7.E − 02	2.E + 01
^{63}Ni[*]	6.E + 04	2.E + 06	6.E + 02	2.E + 04	6.E + 01	2.E + 03
^{103}Pd[*]	9.E + 04	2.E + 06	9.E + 02	2.E + 04	9.E + 01	2.E + 03
^{210}Po[*]	6.E + 01	2.E + 03	6.E − 01	2.E + 01	6.E − 02	2.E + 00
^{106}Ru (^{106}Rh)[*]	3.E + 02	8.E + 03	3.E + 00	8.E + 01	3.E − 01	8.E + 00
^{204}Tl[*]	2.E + 04	5.E + 05	2.E + 02	5.E + 03	2.E + 01	5.E + 02

[a] The primary values to be used are given in TBq. Curie values are provided for convenience and are rounded after conversion.

[b] Criticality and safeguards issues will need to be considered for multiples of D.

[*] These radionuclides are very unlikely to be used in individual sealed radioactive sources with activities that would place them within Categories 1, 2 or 3.

TABLE 6. A/D VALUES FOR REACTOR FUEL

Isotope	Mass fraction	Mass per assembly (kg)	Activity per assembly (TBq)	D value (TBq)	A/D
^{234}U	0.000 336	0.22	0.051	0.1	0.5
^{235}U	0.04	26	0.002 1	0.000 08	30
^{238}U	0.958	630	0.007 8	UL	0

for ^{234}U and ^{235}U are based "on consideration of the criticality limit." This would not apply because the isotopic content and geometry of the uranium as contained in fuel bundles are such that criticality is not possible. Since ^{234}U and ^{235}U have otherwise similar properties to ^{238}U (alpha and low-energy photon emitting, low specific activity), it is likely that the non-criticality D value would be similar or the same (i.e. 'unlimited quantity'), making the A/D value effectively zero.

5.23. Categorization should also take into account radioactive decay and aggregation.

Radioactive decay

5.24. The A/D ratio will decline over time owing to radioactive decay. The regulatory body may take this into account in its regulatory practices.

5.25. For example, a ^{60}Co radioactive source might have an activity of 56 TBq when it is first used in a device. Calculating the A/D ratio, 56 TBq/0.03 TBq = 1867. Therefore, the radioactive source is initially assigned to Category 1. Cobalt-60 has a half-life of 5.2714 years. After three half-lives (approximately 15 years), the ^{60}Co radioactive source has decayed and now has an activity of 7 TBq, and the A/D ratio becomes 7 TBq/0.03 TBq = 233.33, corresponding to an assignment to Category 2. The regulatory body may choose to require this radioactive source to be assigned to Category 1 (based on its original activity ratio) or to allow this radioactive source to be reassigned to Category 2 (based on its current activity ratio). The regulatory body should clearly indicate in its regulations which approach is to be followed.

Aggregation of radioactive material

5.26. There are situations in which multiple items containing radioactive material are in close proximity, such as during manufacturing processes or in storage facilities. Radioactive material should be considered collocated or aggregated if breaching a common physical security barrier (e.g. a locked door at the entrance to a storage room) would allow access to the radioactive material or devices containing the radioactive material.

5.27. In such circumstances where multiple items containing radioactive material are in close proximity, the regulatory body should require operators to aggregate the activity of the radioactive material for the purpose of categorization. In situations of this type, the summed activity of the radionuclide is divided by the appropriate D value and the calculated ratio A/D is compared with the ratios of

A/D given in Table 4, thus allowing the set of radionuclides to be categorized on the basis of activity. If radioactive material composed of various radionuclides is aggregated, then the sum of the ratios A/D should be used in determining the category, in accordance with the formula:

$$\text{Aggregate}\,\frac{A}{D} = \sum_n \frac{\sum_i A_{i,n}}{D_n}$$

where

$A_{i,n}$ = activity of each individual material i of radionuclide n;
D_n = D value for radionuclide n.

Additional information on the aggregation of radioactive material can be found in Ref. [3].

5.28. As an example, in a hospital where brachytherapy is performed, multiple sources may be stored together in a secure room. If all of these radioactive sources can be accessed through a single entry point, they should be aggregated in order to determine their category. For 100 ^{226}Ra radioactive sources (0.001 TBq each), 30 ^{137}Cs radioactive sources (0.02 TBq each) and 10 ^{192}Ir radioactive sources (0.22 TBq), the category would be calculated as follows.

5.29. To determine the category of these collocated sources, the first step is to determine the category of each set of radioactive sources of the same radionuclide. For ^{226}Ra, A/D = (100 × 0.001)/0.04 = 2.5, so these radioactive sources are assigned to Category 3. For ^{137}Cs, A/D = (30 × 0.02)/0.1 = 6, so these radioactive sources are assigned to Category 3. For ^{192}Ir, A/D = (10 × 0.22)/0.08 = 27.5, so these radioactive sources are assigned to Category 2. Because various radionuclides are to be stored together in one secure location, they should be aggregated to obtain a total A/D ratio of 2.5 + 6 + 27.5 = 36. As a result, the aggregate of all the radioactive material to be collocated is assigned to Category 2.

Assigning security levels

5.30. Once radioactive material has been categorized, the next step is to assign security levels to the radioactive material. For example, as a default arrangement, the regulatory body could use the categorization in Table 4 or 5 to assign one of three security levels to the radioactive material. On that basis, Category 1 radioactive material should be assigned to security level A; Category 2 radioactive material to security level B; and Category 3 radioactive material to security level C.

5.31. IAEA Safety Standards Series No. GSR Part 3, Radiation Protection and Safety of Radiation Sources: International Basic Safety Standards [16] includes general requirements for the control of radioactive sources. Those control measures provide a sufficient level of security for radioactive material in Categories 4 and 5. However, the regulatory body, taking into account relevant threat information, may wish to enhance the security of radioactive material in Categories 4 and 5 in appropriate circumstances.

5.32. The approach described in the previous two paragraphs is summarized in Table 7.

Additional considerations for adjusting security levels

5.33. Assigning radioactive material to security levels based on category and practice can be used as a default method. However, on a case by case basis, additional factors specific to the radioactive material and the manner and location in which it is used may lead to an adjustment in the default security level assigned to particular material. In some cases, these factors may lead to the exclusion of certain radioactive material from security requirements altogether.

5.34. The regulatory body may choose to adjust security levels for particular types of radioactive material and include these adjustments in its regulations. For example, it may assign certain types of well logging gauges to security level B, regardless of the security level that would be assigned based on the types of calculation performed in paras 5.30 and 5.31.

5.35. The regulatory body may also allow the operator to propose an adjustment in the security level of its radioactive material, based on specified criteria. In the latter case, the operator would be responsible for seeking approval from the regulatory body for such adjustments.

5.36. A range of specific situations in which the regulatory body may choose to adjust the default security level are described in the following subsections.

Mobile, portable and remote radioactive material

5.37. Radioactive material used in field applications (e.g. industrial radiography and well logging) is typically contained in small devices designed for portability. These devices are frequently transported between job sites and are often used in remote locations. The ease with which these devices can be handled and concealed and their presence in vehicles outside secured facilities makes them vulnerable to unauthorized removal.

5.38. Special consideration should also be given to radioactive material in transport that is incidental to use. For example, an industrial radiography device may be transported to various work sites on a daily basis, which could increase

TABLE 7. DEFAULT SECURITY LEVELS FOR COMMON PRACTICES BY CATEGORY

Category	A/D	Practice/equipment	Security level
1	$A/D \geq 1000$	Radioisotope thermoelectric generators Irradiators Teletherapy Fixed multibeam teletherapy (gamma knife)	A
2	$1000 > A/D \geq 10$	Industrial gamma radiography High/medium dose rate brachytherapy	B
3	$10 > A/D \geq 1$	Fixed industrial gauges that incorporate high activity sources Well logging gauges	C
4	$1 > A/D \geq 0.01$	Low dose rate brachytherapy (except eye plaques and permanent implant sources) Industrial gauges that do not incorporate high activity sources Bone densitometers containing a radioactive source Static eliminators	Apply measures as described in GSR Part 3 [16]
5	$0.01 > A/D$ and $A >$ exempt	Low dose rate brachytherapy eye plaques and permanent implant sources X ray fluorescence devices which contain radioactive material Electron capture devices Mossbauer spectrometry Positron emission tomography check sources	

its vulnerability. Detailed guidance on security in the transport of radioactive material is given in Ref. [11].

5.39. Recognizing that security measures for radioactive material used in a facility may not be practical for application to radioactive material used in the field, additional or alternative security measures may be required. Examples of detection and delay measures for radioactive material used in the field and assigned to security levels B and C are provided in Section 6.

Increased threat

5.40. An increased threat may warrant adjustment of the security level assigned to radioactive material, taking into account all other attributes of the material (e.g. attractiveness or vulnerability). Alternatively, temporary increases in threat may be addressed by requiring the operator to ensure that security measures can be strengthened to address such circumstances (see para. 3.75).

Short half-life radionuclides

5.41. Some fields such as nuclear medicine use radionuclides with short half-lives. Examples of such radionuclides include ^{99m}Tc and ^{18}F, used in radiological diagnostics, and ^{131}I, used in radiotherapy. The regulatory body may conclude that such radioactive material is of low security concern because it is likely to decay before it can be used in a malicious act. Furthermore, even if acquired for malicious purposes, the material would quickly decay below levels which would be harmful. The regulatory body may consider determining a period of time (such as ten days or less), after which radioactive material poses a lower security risk due to radioactive decay and can be adequately secured through assignment to a lower security level or through the application of general requirements for the control of radioactive sources.

Long half-life radionuclides

5.42. Large amounts of long half-life radionuclides might be found in naturally occurring radioactive material (NORM) in concentrations which are too small for realizing malicious acts. Some radionuclides may be unattractive to an adversary due to their low specific activity or low-energy radiation emission. For example, 37 GBq (1 Ci) of depleted uranium has a mass of approximately 2000 kg. For such cases, the regulatory body may choose to reduce the security level, since it would not be practicable to use this material in a malicious act.

Ease of handling

5.43. Radioactive material that can be easily handled or is easily accessible may be attractive to an adversary, because the adversary would be less likely to receive a high radiation dose and the radioactive material is more easily moved. An example of this is a radioactive source inside a self-shielded portable device.

Large volumes of activated radioactive material or contaminated objects

5.44. Legacy sites as well as operating facilities may contain activated or contaminated components and structures which are not normally considered subject to specific security requirements during the operating lifetime of a reactor, hot cell or accelerator. Examples include various metallic parts from steam generators and dryers, turbine rotors, reactor vessels and vessel heads, reactor coolant pumps, and shielding blocks.

5.45. Due to their size and weight, large activated or contaminated components are not easily moved without cranes, rigging and heavy equipment. In addition, these large components are not easily concealed during loading or when they are in motion, and the amount of time required to remove these large components is such that it is reasonable to expect that the operator would detect these activities. Further, if such a large component were to be removed, it would be very difficult to use in a malicious act.

5.46. The regulatory body could choose to exempt these components from security requirements or to reduce their security level if the operator demonstrates that such an adjustment is warranted. The regulatory body should strike a balance between the advantages of keeping the components under regulatory control and the relatively low risk posed by the components, in accordance with a graded approach.

Location of radioactive material

5.47. For radioactive material located in a densely populated area where its use in a malicious act could be a greater security concern than in a less populated area, the regulatory body may consider increasing the assigned security level above the default. One example is the use of radioactive material for cancer treatment in a hospital located in a densely populated city. In this case, an increase in the security level assigned to that material may be warranted. Considerations affecting response times, such as the distance between the facility where radioactive material is located and the place where the local response force is stationed, could also factor into a decision to adjust the default security level.

Radioactive waste

5.48. In principle, security levels can be assigned to radioactive waste in the same manner as described in paras 5.30–5.32 for other radioactive material. However, several considerations may lead to adjustments in the assignment of security levels to radioactive waste.

5.49. The State or the regulatory body may choose to reduce the default security level to reflect the lesser attractiveness to potential adversaries of some forms of radioactive waste compared with other radioactive material of comparable radioactivity. Attributes leading to lower attractiveness include:

— *Recoverability:* Radioactive waste might be contained in a solid matrix (e.g. a concrete block), making it difficult to recover.
— *Susceptibility to dispersal:* Radioactive waste contained within a solid matrix is not readily susceptible to dispersal.
— *Feasibility of transport:* The weight of certain types of radioactive waste containers makes them difficult to transport because doing so is time consuming and requires the use of heavy equipment.

5.50. In addition, the default security level may be reduced to reflect the lesser vulnerability of radioactive waste in certain storage or disposal locations. Depending on the State's regulatory requirements and infrastructure, radioactive waste could be located in short term storage at an operator's facility, in long term storage at a dedicated (centralized) storage facility or in a disposal facility. Within a disposal facility, radioactive waste could be located in either of two primary areas: an operations area that is actively receiving, sorting and emplacing the radioactive waste; or a disposal area where radioactive waste has been disposed of, such as a borehole.

5.51. Radioactive waste in short term storage at an operations facility, in long term storage at a dedicated (centralized) storage facility or in the operations area of a disposal facility could be assigned to the same security level as other radioactive material of comparable activity. The security level could also be reduced based on waste form or package, as discussed in para 5.49.

5.52. Radioactive waste emplaced in a disposal area is often less accessible to adversaries than radioactive waste present in other locations because the disposal area typically includes limited access points and one or more physical barriers. For example, an adversary attempting to remove radioactive waste from a repository or borehole would likely be detected before completing the removal due to the scale, visibility and time required for the attempted recovery operations. For this reason, the regulatory body may establish specific security requirements for disposal areas that differ from those that would otherwise be applicable.

5.53. The default security level may also be adjusted based on the potential for sabotage resulting in radiological contamination. While radioactive waste located in a disposal area might not be attractive to adversaries for unauthorized removal, it might have some attractiveness as a sabotage target. In this case, the regulatory body could establish specific security requirements to protect disposal areas against sabotage.

Other considerations

5.54. Additional factors which may warrant an adjustment of the security level assigned to radioactive material include the perceived economic value of the radioactive material or associated device and the presence of other hazardous material within the facility where radioactive material is located.

STEP 3: IMPLEMENT A REGULATORY APPROACH

5.55. As described in Section 3, there are three regulatory approaches that the regulatory body may use to establish security requirements for radioactive material: prescriptive, performance based and combined.

5.56. Regardless of the approach used, regulatory requirements for the security of radioactive material should address each of the following questions:

— What is the operator required to protect?
— What is the operator required to protect against?
— What degree of protection is considered adequate?
— What are the security measures the operator is required to implement?
— What are the security management measures the operator is required to implement?

5.57. As summarized in Table 8, regulations based on the prescriptive and performance based approaches address some of these questions in the same way and other questions in different ways. How regulations based on the combined

TABLE 8. COMPARISON OF REGULATIONS BASED ON PRESCRIPTIVE
AND PERFORMANCE BASED APPROACHES

Question	Prescriptive approach	Performance based approach
What is the operator required to protect?	Specified radioactive material and associated facilities	Specified radioactive material and associated facilities
What is the operator required to protect against?	The threat, as used by the regulatory body to develop prescriptive security requirements (complete threat information not generally provided to the operator)	The threat, as provided by the regulatory body to the operator for use in designing its security system
What degree of protection is considered adequate?	Security level A, B or C goal and sub-goals, as applicable to the material	Security level A, B or C goal and sub-goals, as applicable to the material
What are the security measures the operator is required to implement?	The security measures required by the regulatory body based on its determination that they will generally be sufficient to meet the applicable security level A, B or C goal and sub-goals against the threat	The security measures proposed by the operator and accepted by the regulatory body as sufficient to meet the applicable security level A, B or C goal and sub-goals against the threat
What are the security management measures the operator is required to implement?	The security management measures required by the regulatory body	The security management measures proposed by the operator and accepted by the regulatory body as sufficient to meet goals for all security levels

approach address these questions will depend on how the regulatory body chooses
to combine the two approaches. Each of these questions is further explained in
paras 5.58–5.68.

What is the operator required to protect?

5.58. Regardless of approach, regulations should specify the radioactive material
and threshold activities above which the operator is required to implement

security measures. Security regulations for radioactive material typically apply to all radionuclides determined by the regulatory body to represent a security concern based on their potential to cause harmful radiological consequences if used in a malicious act.

5.59. As discussed in para. 5.31, the regulatory body may choose to exclude Category 4 and 5 radioactive material from specific security requirements and instead mandate the application of measures described in GSR Part 3 [16]. In addition, certain material could be excluded from security requirements based on the additional considerations for assigning security levels discussed in paras 5.33–5.54.

What is the operator required to protect against?

5.60. Regulations based on either approach should require the operator to protect against the threat defined by the threat assessment and the DBT or RTS, as described in Section 3. The regulatory body should apply threat information in a manner consistent with the regulatory approach chosen.

5.61. If a prescriptive approach is selected, the regulatory body should adopt regulations that specify a set of required security measures for detection, delay and response. This set of security measures should be sufficient, if appropriately implemented, to meet the applicable security goals and sub-goals as determined by the regulatory body. When this approach is used, the regulatory body does not typically convey the threat information to operators, except in very general terms.

5.62. If a performance based approach is selected, the regulatory body should adopt regulations which require the operator to design and implement a security system that is sufficient to meet the applicable security goal and sub-goals, depending on the type of material to be protected and in accordance with a graded approach. When this approach is used, the regulatory body directly shares relevant threat information with operators, subject to stringent information protection requirements.

What degree of protection is considered adequate?

5.63. Regulations developed based on either a prescriptive or a performance based approach should require the operator to meet applicable security goals and sub-goals.

5.64. Prescriptive regulations would require the operator to implement required security measures in a manner that is determined by the regulator to meet the applicable security goals and sub-goals, described in Table 3. For example, prescriptive regulations for security level A radioactive material should require the operator to implement specified security measures in a manner that provides a very high level of confidence that the security system will prevent unauthorized removal of radioactive material.

5.65. Performance based regulations should require the operator to design and implement a security system that is sufficient to meet the applicable security goals and sub-goals, given the threat defined by the threat assessment and, as applicable, the DBT or RTS, and communicated to the operator by the regulatory body.

What are the security measures the operator is required to implement?

5.66. As described in the previous sections, regulations developed based on a prescriptive approach require the operator to implement specified security measures. However, because of the wide variation in facilities and activities involving the use or storage of radioactive material, regulations should give the operator appropriate discretion in implementing the required measures. For example, the regulations may require the operator to implement electronic intrusion detection systems, while leaving the operator flexibility to make such choices as which particular technologies to deploy (e.g. balanced magnetic switch, passive infrared sensors) and how to configure the chosen technologies.

5.67. Performance based regulations require the operator to design and implement a security system consisting of security measures which, implemented together, protect the radioactive material against the threat.

What are the security management measures the operator is required to implement?

5.68. Regulations based on either approach should specify security management measures that the operator is required to implement, addressing at a minimum:

— Access control;
— Trustworthiness;
— Information protection;
— A security plan;
— Training and qualification;
— Accounting;

— Inventory;
— Security system evaluation;
— Nuclear security event reporting and post-event reporting.

6. GUIDANCE ON THE CONTENT OF REGULATIONS

6.1.   This section provides guidance on the content of regulations for the regulatory approaches described in Section 5. The guidance provided in this section on the prescriptive approach includes specific security measures. More general guidance is provided on the performance based and combined approaches.

PRESCRIPTIVE APPROACH

6.2.   With a prescriptive approach, the regulatory body chooses to establish regulations which specify the security measures that operators are required to have in place in order to meet the security sub-goals described in Table 3. Tables 9, 10 and 11 provide suggested measures for detection, delay and response for security levels A, B and C, respectively, applicable to radioactive material in use or in storage. Table 10 also includes specific security measures for portable devices used in the field. Table 12 identifies security management measures for all three security levels. The measures are reproduced and discussed in detail after each corresponding table. This text in this section is intended primarily to clarify the tables, but could also be incorporated selectively in regulations or guidance.

6.3.   The regulatory body should require that the operator implement the measures in a manner that meets the applicable security sub-goal.

Security level A measures

6.4.   The goal of nuclear security for radioactive material assigned to security level A is to provide a high level of protection of radioactive material against unauthorized removal. If unauthorized access or unauthorized removal is attempted, detection and assessment should occur early enough and delay needs to impede the adversary long enough to enable response personnel to respond in time and with sufficient resources to interrupt the adversary and prevent the radioactive material from being removed.

6.5. The measures described in Table 9 and in the following subsections should be required to achieve the goal above for protecting material assigned to security level A.

TABLE 9.   DETECTION, DELAY AND RESPONSE MEASURES FOR SECURITY LEVEL A

(Goal: Provide a high level of protection of radioactive material against unauthorized removal)

Security function	Security sub-goal	Security measures
Detection	Provide immediate detection of unauthorized access to locations in which radioactive material is present Provide immediate detection of attempted unauthorized removal of radioactive material, including removal by an insider	Electronic intrusion detection system and/or continuous surveillance by operator personnel
	Provide immediate assessment of detection	Remote video monitoring and/or direct observation by operator or response personnel
	Provide a means to detect loss through verification	Daily verification through such measures as physical checks, video monitoring, tamper indicating devices
Delay	Furnish sufficient delay to provide a high level of protection against unauthorized removal of radioactive material	System of at least two layers of barriers (e.g. walls, cages)
Response	Provide immediate communication to response personnel	Rapid, dependable, diverse means of communication such as telephones, mobile phones and/or radios
	Provide for immediate response with sufficient resources to interrupt and prevent the unauthorized removal of radioactive material	Arrangements with a designated response force, including provision for sufficient personnel, equipment and training, documented in a response plan

Security sub-goals:	Provide immediate detection of unauthorized access to locations in which radioactive material is present.
	Provide immediate detection of attempted unauthorized removal of radioactive material, including removal by an insider.
Security measures:	Electronic intrusion detection system and/or continuous surveillance by operator personnel.

6.6.   Electronic sensors linked to an alarm or continuous visual surveillance by operator personnel indicate either unauthorized access to the location in which radioactive material is present (see paras 4.5, 4.6) or attempted unauthorized removal of radioactive material. Care should be taken to ensure that such measures cannot be bypassed. For radioactive material in use, measures should detect unauthorized access to the secured locations where the radioactive material is used. For radioactive material in storage, measures should detect unauthorized access to the locked room or other location where the radioactive material is stored.

Security sub-goal:	Provide immediate assessment of detection.
Security measure:	Remote video monitoring and/or direct observation by operator or response personnel.

6.7.   Once an alarm has been triggered, the cause of the alarm should be assessed immediately. Assessment can be performed by operator personnel at the location where radioactive material is present through remote video monitoring (e.g. at a central alarm station), or by persons immediately deployed to investigate the cause of the alarm. While video monitoring is an effective assessment measure, it is not a reliable detection measure and should not be used for this purpose.

Security sub-goal:	Provide a means to detect loss through verification.
Security measures:	Daily verification through such measures as physical checks, video monitoring, tamper indicating devices.

6.8.   Daily verification should consist of measures to ensure that the radioactive material is present and neither the radioactive material nor the device in which it is contained has been tampered with. Such measures could include physical checks that the radioactive material is in place, remote video monitoring, verification of seals or other tamper indicating devices, and measurements of radiation or other physical phenomena that would provide an assurance that the radioactive material is present. For radioactive material in use, verifying that the corresponding device is intact and functional may be sufficient.

Delay

Security sub-goal: Furnish sufficient delay to provide a high level of protection against unauthorized removal of radioactive material.

Security measures: System of at least two layers of barriers (e.g. walls, cages).

6.9. A balanced system comprising at least two barriers should separate the radioactive material from unauthorized personnel. This system should provide sufficient delay following detection to enable response personnel to intercede before an adversary could remove the radioactive material or the device in which it is contained. For radioactive material in use, such measures could include maintaining the material in a locked device in a secured area to separate the device from unauthorized personnel. For radioactive material in storage, such measures could include a locked and fixed container or a device holding the radioactive material in a locked storage room.

Response

Security sub-goal: Provide immediate communication to response personnel.

Security measures: Rapid, dependable, diverse means of communication such as telephones, mobile phones and/or radios.

6.10. If the assessment confirms that unauthorized access or attempted unauthorized removal has occurred, operator personnel should immediately notify response personnel. Accordingly, such personnel should be equipped with at least two separate means of communication, such as telephones, mobile phones and/or radios. Where detection and assessment are performed directly by operator personnel, the location should be equipped with fixed or mobile duress buttons.

Security sub-goal: Provide for immediate response with sufficient resources to interrupt and prevent the unauthorized removal of radioactive material.

Security measures: Arrangements with a designated response force, including provision for sufficient personnel, equipment and training, documented in a response plan.

6.11. In most cases, the operator will not be capable of providing its own response and instead will rely on an external response force, typically law enforcement personnel. The State should identify the entity expected to provide such a response. The operator should be required to establish arrangements with the designated response force to ensure immediate deployment of response personnel in response to an alarm. The regulatory body should facilitate the establishment of these arrangements.

6.12. Responders should arrive, once notified, within a time period shorter than the time needed for an adversary to breach the barriers and perform the tasks needed to remove the radioactive material. The response team should be of sufficient size and capability to defeat the adversary. The operator's response arrangements should be documented in the security plan and/or response plan, as further discussed in paras 6.47–6.52 and 6.60–6.63.

Security level B measures

6.13. The goal of nuclear security for material assigned to security level B is to provide an intermediate level of protection of radioactive material against unauthorized removal. If unauthorized access or unauthorized removal were to be attempted, the response should be initiated immediately upon detection and assessment of the intrusion, but in contrast to security level A, the response does not need to be required to arrive in time to prevent the radioactive material from being removed.

6.14. The measures described in Table 10 and in the following subsections should be required to achieve the goal above for protecting radioactive material assigned to security level B. Because security level B radioactive material is often used in portable devices deployed in the field, which cannot be protected in the same manner as radioactive material used or stored in fixed locations, Table 10 and the accompanying text also include specific security measures that may be additionally or alternatively required.

TABLE 10. DETECTION, DELAY AND RESPONSE MEASURES FOR SECURITY LEVEL B

(Goal: Provide an intermediate level of protection of radioactive material against unauthorized removal)

Security function	Security sub-goal	Security measures (radioactive material in use and storage)	Security measures (portable devices containing radioactive material when used in the field)
Detection	Provide immediate detection of unauthorized access to locations where radioactive material is present	Electronic intrusion detection system and/or continuous surveillance by operator personnel	Visual observation by two operator personnel
	Provide detection of attempted unauthorized removal of radioactive material	Tamper detection equipment and/or periodic checks by operator personnel	Visual observation by two operator personnel
	Provide immediate assessment of detection	Remote video monitoring or direct observation by operator and/or response personnel	Observation by operator personnel
	Provide a means to detect loss through verification	Weekly verification through measures such as physical checks and tamper detection equipment	Daily checks after field use
Delay	Furnish sufficient delay to provide an intermediate level of protection against unauthorized removal of radioactive material	System of two layers of barriers (e.g. walls, cages)	Means of affixing the device to a stationary object, if possible

TABLE 10. DETECTION, DELAY AND RESPONSE MEASURES FOR SECURITY LEVEL B

(Goal: Provide an intermediate level of protection of radioactive material against unauthorized removal) (cont.)

Security function	Security sub-goal	Security measures (radioactive material in use and storage)	Security measures (portable devices containing radioactive material when used in the field)
Response	Provide immediate communication to response personnel	Rapid, dependable means of communication such as telephones, mobile phones and/or radios	Two persons, each equipped with an independent mobile communication device
	Provide immediate initiation of response to interrupt unauthorized removal	Equipment and procedures to immediately initiate response	Advance notification to local response force before deployment and immediate communication after detection

Detection

Security sub-goal: Provide immediate detection of unauthorized access to locations where radioactive material is present.

Security measures: *For fixed facilities:* Electronic intrusion detection system and/or continuous surveillance by operator personnel.

For portable devices: Visual observation by two operator personnel.

6.15. Electronic sensors linked to an alarm or continuous visual surveillance by operator personnel can be used to indicate unauthorized access to the location of radioactive material.

6.16. Visual observation by two operator personnel can be used for immediate detection of unauthorized access to radioactive material contained in portable or mobile devices.

Security sub-goal: Provide detection of attempted unauthorized removal of radioactive material.

Security measures: *For fixed facilities:* Tamper detection equipment and/or periodic checks by operator personnel.

For portable devices: Visual observation by two operator personnel.

6.17. Tamper detection equipment or visual surveillance by operator personnel made during periodic checks can be used to detect attempted unauthorized removal of radioactive material.

6.18. Visual observation by two operator personnel or radiation monitoring can be used for immediate detection of unauthorized removal of radioactive material contained in portable or mobile devices.

Security sub-goal: Provide immediate assessment of detection.

Security measures: *For fixed facilities:* Remote video monitoring or direct observation by operator and/or response personnel.

For portable devices: Observation by operator personnel.

6.19. Once an alarm has been triggered, the cause of the alarm should be assessed immediately. For radioactive material in use and storage, alarms can be assessed either through remote video monitoring or through observation by operator or response personnel.

6.20. In the case of portable devices, observation by operator personnel is the only feasible means of assessment.

Security sub-goal: Provide a means to detect loss through verification.

Security measures: *For fixed facilities:* Weekly verification through measures such as physical checks and tamper detection equipment.

For portable devices: Daily checks after field use.

6.21. Weekly verification consists of measures to ensure that the radioactive material is present and neither the radioactive material nor the device in which it is contained has been tampered with. The section on detection for radioactive material assigned to security level A contains some examples of such measures.

6.22. For portable devices, radioactive material should be checked daily after use in the field.

Delay

Security sub-goal: Furnish sufficient delay to provide an intermediate level of protection against unauthorized removal of radioactive material.

Security measures: *For fixed facilities* System of two layers of barriers (e.g. walls, cages).

For portable devices: Means of affixing the device to a stationary object, if possible.

6.23. A balanced system of two barriers should separate radioactive material in use or storage from unauthorized personnel.

6.24. Portable devices should be affixed to a stationary object in order to delay their removal.

Response

Security sub-goal: Provide immediate communication to response personnel.

Security measures: *For fixed facilities:* Rapid, dependable means of communication such as telephones, mobile phones and/or radios.

For portable devices: Two persons, each equipped with an independent mobile communication device.

6.25. If the assessment of a detected event confirms that unauthorized access or attempted unauthorized removal has occurred, response personnel should be immediately notified once the assessment is complete.

6.26. In the case of portable devices used in the field, there should be two operator personnel at the location, each equipped with mobile communication equipment. Each piece of communications equipment should operate independently and be tested in advance to ensure coverage.

Security sub-goal: Provide immediate initiation of response to interrupt unauthorized removal.

Security measures: *For fixed facilities:* Equipment and procedures to immediately initiate response.

            For portable devices: Advance notification to local response force before deployment and immediate communication after detection.

6.27. The operator should establish arrangements to ensure immediate deployment of response personnel to interrupt an adversary action following the detection and assessment of an alarm.

6.28. Operators using portable devices in the field should provide advance notification of their presence to the local response force before deploying the devices and communicate with the response force immediately following the detection and assessment of an attempted unauthorized removal.

Security level C measures

6.29. The goal of nuclear security for material assigned to security level C is to provide a baseline level of protection of radioactive material against unauthorized removal. To the extent appropriate and feasible, the regulatory body may choose to require security level B security measures for portable devices containing security level C radioactive material when used in the field.

6.30. The measures described in Table 11 and in the following subsections should be required to achieve the goal above for protecting radioactive material assigned to security level C.

TABLE 11. DETECTION, DELAY AND RESPONSE MEASURES FOR SECURITY LEVEL C

(Goal: Provide a baseline level of protection of radioactive material against unauthorized removal)

Security function	Security sub-goal	Security measures
Detection	Provide detection of unauthorized removal of radioactive material	Observation by operator personnel
	Provide a means to detect loss through verification	Monthly verification through such measures as physical checks, tamper detection equipment
Delay	Furnish sufficient delay to provide a baseline level of protection against unauthorized removal of radioactive material	One barrier (e.g. cage, source housing) and/or presence of operator personnel
Response	Provide prompt communication to response personnel	Rapid, dependable means of communication such as telephones, mobile phones and/or radios
	Implement appropriate action in the event of unauthorized removal of radioactive material	Procedures for identifying necessary actions in accordance with response plan

Detection

Security sub-goal: Provide detection of unauthorized removal of radioactive material.

Security measures: Observation by operator personnel.

6.31. Operator personnel should be trained to be vigilant when unauthorized persons are being escorted in the facility.

Security sub-goal: Provide a means to detect loss through verification.

Security measures: Monthly verification through such measures as physical checks, tamper detection equipment devices.

6.32. Monthly verification consists of measures to ensure that the radioactive material is present and neither the radioactive material nor the device in which it is contained has been tampered with. Such measures could include physical checks that the radioactive material is in place, and verification of seals or other tamper detection equipment. If tamper detection or a physical check indicates that radioactive material might be missing, the situation should be assessed immediately to determine whether an unauthorized removal has occurred. The section on protection of radioactive material assigned to security level A contains some examples of such measures.

Delay

Security sub-goal: Furnish sufficient delay to provide a baseline level of protection against unauthorized removal of radioactive material.

Security measures: One barrier (e.g. cage, source housing) and/or presence of operator personnel.

6.33. At least one physical barrier should separate the radioactive material from unauthorized personnel. Such measures could include the radioactive source housing or use of the radioactive material in a secured area. The presence of operator personnel could also be used to delay unauthorized access to radioactive material.

Response

Security sub-goal: Provide prompt communication to response personnel.

Security measures: Rapid, dependable means of communication such as telephones, mobile phones and/or radios.

6.34. If the assessment of a detected event confirms that unauthorized access or attempted unauthorized removal has occurred, response personnel should be promptly notified.

Security sub-goal: Implement appropriate action in the event of unauthorized removal of radioactive material.

Security measures: Procedures for identifying necessary actions in accordance with response plan.

6.35. Regulatory procedures should ensure that any suspected unauthorized removal or loss of radioactive material is assessed and, if confirmed, reported to the appropriate authority without delay. This should be followed by an effort to locate and recover the radioactive material and investigate the circumstances leading to the event.

Security management measures

6.36. Security sub-goals and measures for security management are the same for security levels A, B and C. However, the operator should apply a graded approach in implementing the security measures. In some cases, the paragraphs that follow provide specific guidance on how the graded approach should be applied. In other cases, the specifics of implementation are to be undertaken at the discretion of the regulatory body and/or operator.

6.37. The measures described in Table 12 and in the following subsections should be required to achieve the goal above for protecting radioactive material.

TABLE 12. SECURITY MANAGEMENT MEASURES

Security sub-goal	Security measures
Establish a process for granting individuals authorized unescorted access to radioactive material and/or access to sensitive information	Procedures for determining the individuals who need access, verifying that such individuals are trustworthy and reliable and have received necessary training, authorizing access, withdrawing access as appropriate and maintaining documentation
Ensure trustworthiness and reliability of authorized individuals	Background checks for all personnel authorized for unescorted access to radioactive material and/or for access to sensitive information
Provide access controls that effectively restrict unescorted access to radioactive material to authorized persons only	Identification and verification measures

TABLE 12. SECURITY MANAGEMENT MEASURES (cont.)

Security sub-goal	Security measures
Identify and protect sensitive information	Procedures to identify sensitive information and protect it from unauthorized disclosure
Provide a security plan	A security plan which addresses required topics, is submitted or made available to the regulatory body and is periodically exercised, evaluated and revised, as appropriate
Ensure training and qualification of individuals with security responsibilities	Assessment of necessary knowledge, skills and abilities; provision of corresponding training; procedures for documenting and updating training
Conduct accounting and inventory of radioactive material	Procedures and documentation for verifying presence of radioactive material at prescribed intervals; establishment and maintenance of a radioactive material inventory
Conduct evaluation for compliance and effectiveness, including performance testing	Process for verifying that all applicable security requirements are met and for assessing the effectiveness of the security system, employing performance tests as appropriate
Establish a capability to manage and report nuclear security events	Response plan addressing security related scenarios and procedures for timely reporting of nuclear security events

Security sub-goal: Establish a process for granting individuals authorized unescorted access to radioactive material and/or access to sensitive information.

Security measures: Procedures for determining the individuals who need access, verifying that such individuals are trustworthy and reliable and have received necessary training, authorizing access, withdrawing access as appropriate and maintaining documentation.

6.38. The regulatory body should require operators to limit unescorted access to radioactive material and access to sensitive information to those individuals with a demonstrated need for such access in the performance of their jobs, whose trustworthiness has been verified, and who have received necessary security training. The process for granting access authorization to such individuals should include the following steps:

(a)   Determining that an individual needs such access in order to discharge his or her responsibilities;
(b)   Obtaining verification that the individual is trustworthy and reliable (see paras 6.39 and 6.40);
(c)   Obtaining verification that the individual has received necessary security training for the access authorization in question (see paras 6.41–6.44);
(d)   Authorizing access based on the determination of a need for access and the verifications obtained in steps (b) and (c);
(e)   Withdrawing access as appropriate, for example when an individual's responsibilities change or when employment is terminated;
(f)   Maintaining current documentation of the results of this process and providing it to those responsible for access control.

Security sub-goal: Ensure trustworthiness and reliability of authorized individuals.

Security measures: Background checks for all personnel authorized for unescorted access to radioactive material and/or for access to sensitive information.

6.39. An individual's trustworthiness should be assessed through a satisfactory background check before that individual is allowed unescorted access to radioactive material or locations where radioactive material is used or stored and before that person is allowed access to any related sensitive information. The nature and depth of background checks should be proportionate to the security level of the radioactive material (i.e. more thorough background checks should be performed for radioactive material assigned to a higher security level) and in accordance with the State's regulations or as determined by the regulatory body. At a minimum, background checks should confirm identity and verify references to determine the trustworthiness and reliability of the individual being assessed. The checks could also include disclosure of criminal conduct. The process should be periodically reviewed and supported through ongoing monitoring by supervisors and managers to ensure that personnel at all levels continue to act responsibly and reliably and that any concerns, in this context, are made known to the relevant authority. Periodic background checks of employees whose trustworthiness has been previously assessed (e.g. every 5 years) should also be conducted, as long as those employees continue to need unescorted access to radioactive material or locations where radioactive material is used or stored or access to any related sensitive information.

6.40. In many States, the operator will not be authorized to perform, or be capable of performing, background checks and will instead be reliant on law enforcement, the justice ministry or another competent authority to perform such checks at the operator's request. In such cases, the regulatory body should identify the entity responsible for performing background checks within the State's governmental system and facilitate the necessary communications between operators and this entity. The results of background checks should be considered sensitive for both security and privacy reasons and should be protected accordingly.

Security sub-goal: Provide access controls that effectively restrict unescorted access to radioactive material to authorized persons only.

Security measures: Identification and verification measures

6.41. Access control is intended to limit access to locations where radioactive material is present to authorized persons. Access control typically consists of allowing such persons to temporarily disable physical barriers such as a locked door only upon verification of the person's identity and access authorization.[10]

6.42. The identity and authorization of a person seeking access can be verified using such measures as:

— Personal identification number to activate a door control reader;
— A badge system which could also activate an electronic reader;
— A badge exchange scheme at an entry control point;
— Biometric features to activate a door control device.

Upon verification of a person's identity and access authorization, the system allows that person to enter the secured area or location of radioactive material (e.g. by opening a lock).

---

[10] In the context of medical exposure, patients do not need to be 'authorized' since they are escorted to the radioactive source and are under constant surveillance by the medical staff.

6.43. For security level A, a combination of two or more verification measures should be required, for example the use of a swipe card and a personal identification number or the use of a key combined with visual verification of identity by other authorized personnel.

6.44. For security levels B and C, at least one verification measure should be required.

Security sub-goal: Identify and protect sensitive information.

Security measures: Procedures to identify sensitive information and protect it from unauthorized disclosure.

6.45. According to Ref. [12], sensitive information is information, the unauthorized disclosure (or modification, alternation, destruction or denial of use) of which could compromise nuclear security or otherwise assist in the carrying out of a malicious act against a nuclear facility, organization or transport. This definition also applies to radioactive material, associated facilities and associated activities. Such information may include documents, data on computer systems and other media that can be used to identify details of:

— The nuclear security arrangements at a facility;
— The systems, structures and components at a facility;
— The location and details of transport of radioactive material (sources);
— Details of an organization's personnel.

6.46. The regulatory body should require the operator to establish procedures for identifying such information and for protecting it from disclosure during use, storage and transmission. Information security measures are set out in more detail in Ref. [12].

Security sub-goal: Provide a security plan.

Security measures: A security plan which addresses required topics, is submitted or made available to the regulatory body and is periodically exercised, evaluated and revised, as appropriate.

6.47. The operator should be required to develop, implement, exercise, evaluate and revise as necessary a security plan which documents the design, operation and maintenance of the entire security system as well as the implementation of the security management elements of the security system. The security plan both enables operators to demonstrate to the regulatory body their compliance with security requirements and provides relevant information to facility security personnel for the operation, maintenance and continuous improvement of the security system. Appendix II provides an example of topics that a security plan could be required to address.

6.48. Security plans should be submitted or made available to the regulatory body for review as part of the authorization or inspection process. The operator should be required to exercise, evaluate and revise the security plan at least annually to ensure that it reflects the current security system and remains effective. Security plans contain sensitive information and should be managed accordingly.

6.49. The detail contained within a security plan as well as the frequency with which it is exercised, evaluated and revised should be commensurate with the security level of the radioactive material.

Security sub-goal:	Ensure training and qualification of individuals with security responsibilities.
Security measures:	Assessment of necessary knowledge, skills and abilities; provision of corresponding training; procedures for documenting and updating training.

6.50. The operator should be required to establish requirements for qualification of staff with specific security responsibilities. Such qualification requirements should be based on an assessment of the knowledge, skills and attitudes necessary to meet the assigned security responsibilities; should generally include minimum educational qualification and previous experience; and may also include minimum physical qualifications, security clearance requirements and experience or training in the operation of specific security equipment and the implementation of security procedures. The regulatory body should require the operator to assess each individual against the applicable qualification requirements before assigning that individual to a position with security responsibilities, provide necessary training, periodically reassess the competence of such staff to perform their assigned duties (requalification) and provide retraining as appropriate. Such training should include the use of drills and exercises, as appropriate. All staff should receive general security awareness training.

6.51. Training and qualification of all facility personnel should be documented and the records maintained. All training courses and materials should also be regularly reviewed for relevance of content and effectiveness of delivery.

6.52. The extent of training and qualification should depend on the knowledge, skills and abilities needed for security personnel to meet their responsibilities, commensurate with the security level of the operator's radioactive material.

Security sub-goal: Conduct accounting and inventory of radioactive material.

Security measures: Procedures and documentation for verifying presence of radioactive material at prescribed intervals; establishment and maintenance of a radioactive material inventory.

6.53. Detecting the loss of radioactive material through verifying by periodic checking is addressed in paras 6.8, 6.21, 6.22 and 6.32. Accounting for and taking inventory of radioactive material involves the operator maintaining a record indicating the results of each of these periodic checks, including the date and time when the check was performed, the individual who performed the check and the means used to verify the presence of the radioactive material. If the presence of the radioactive material cannot be verified, the regulatory body should require that the operator report to the regulatory body and/or other government authorities, in a manner and within a time prescribed by regulation, and assist as requested in efforts to locate and recover the radioactive material.

6.54. The regulatory body should also require the operator to establish and maintain an inventory of all radioactive material the operator is authorized to possess.

6.55. The regulatory body should require the operator to adjust the inventory to reflect transfers and receipts within a time prescribed by the regulatory body. Annually, or at another more frequent interval as specified by the regulatory body, the operator should verify that the inventory is complete and accurate and adjust the inventory to reflect any discrepancies identified. The regulatory body should require the operator to report these inventory results to the regulatory body for inclusion in the national registry of radioactive material or radioactive sources.

Security sub-goal: Conduct evaluation for compliance and effectiveness, including performance testing.

Security measures: Process for verifying that all applicable security requirements are met and for assessing the effectiveness of the security system, employing performance tests as appropriate.

6.56. Evaluation is a process by which the operator independently verifies that its facility is in compliance with all applicable security requirements and assesses the effectiveness of its security system to identify any weaknesses that should be corrected and any opportunities for continuous improvement. Evaluation helps ensure that the operator's security system is reliably operated and maintained, functions as intended, is effective and meets regulatory requirements.

6.57. Performance tests provide one especially useful means of evaluating elements of the security system in order to determine whether they can actually perform as required by the regulatory body or produce the desired results. Performance testing, which should be integral to the evaluation process, includes the investigation, measurement, validation or verification of one or more of the following:

— Personnel, to verify that they understand the security system, follow procedures and use the system properly and as intended;
— Procedures, to verify that the procedures produce the desired result and that personnel understand and properly follow them;
— Equipment, to verify that equipment functions as intended and is effective.

6.58. The regulatory body should require the operator to develop and implement an evaluation process that includes performance tests, as appropriate.

6.59. The comprehensiveness of the evaluation process used should be commensurate with the security level assigned to the radioactive material.

Security sub-goal: Establish a capability to manage and report nuclear security events.

Security measures: Response plan addressing security related scenarios and procedures for timely reporting of nuclear security events.

6.60. The regulatory body should require the operator to develop a response plan for a range of potential nuclear security events, including:

— A suspected or threatened malicious act;
— A public demonstration which has the potential to threaten the security of sources;
— Unauthorized access to a location in which radioactive material is present;
— Attempted or successful unauthorized removal of radioactive material.

6.61. The operator should develop a response plan addressing these and any other reasonably foreseeable scenarios involving nuclear security events as well as procedures for responding to them. The response plan could be prepared as part of the security plan or as a separate document. External security response forces, as well as emergency response personnel, should be consulted to ensure that their roles and responsibilities are appropriately understood and documented in the response plan and should be provided with adequate radiation protection. The response plan should be exercised at regular intervals (at least annually) and modified as necessary to address identified weaknesses. The response plan should be coordinated with the radiological emergency plan.

6.62. The response plan should include procedures for reporting of nuclear security events to the regulatory body, response forces, emergency response organizations and others, as appropriate, within a time frame required by the regulatory body. This time frame should be commensurate with the significance of the event, based on a graded approach. Events that may be reported include:

— Discrepancies in inventory data;
— Unauthorized access to radioactive material;
— Suspected or actual unauthorized removal of radioactive material;
— Unauthorized access to sensitive information;
— Failure or loss of security systems that are essential to the protection of radioactive material;
— Other malicious acts that threaten authorized activities.

6.63. The level of detail contained within a response plan as well as the frequency with which it is exercised, evaluated and revised should be commensurate with the security level assigned to the radioactive material.

PERFORMANCE BASED APPROACH

6.64. The regulatory body may choose to specify the use of a performance based approach in which operators are required to meet applicable security sub-goals, as set by the regulatory body. A State's selection of this approach will usually depend on the availability of security expertise to the regulatory body and the operator. A performance based approach functions most effectively when operators have professional advisers and expertise available to design and implement the necessary security measures and have demonstrated a sustained record of consistency and compliance. The regulatory body should ensure that the approved measures are clearly documented (e.g. within a security plan which is reviewed and updated periodically and assessed at appropriate intervals).

6.65. If a performance based approach is selected, a State will need to use the national threat assessment as the basis for the approach, and could also choose to develop a DBT or RTS. The regulatory body should further specify security goals and sub-goals for the security levels of radioactive material for which the performance based approach applies. The security sub-goals should usually be stated in terms of required system effectiveness, as discussed in Section 3.

6.66. The operators should design a security system that meets the applicable security goals and sub-goals by evaluating the security system against the applicable threat information. The operator should use either the evaluation approach described in Section 3 or another methodology, as determined by the regulatory body. The results evaluation (performed using a vulnerability assessment or other methodology) would also be used to demonstrate that the resulting security system does, in fact, meet the applicable security goals and sub-goals.

6.67. The set of security measures developed by applying the performance based approach would not necessarily correspond to the security measures that would be required by the prescriptive approach based on Tables 9–11 for a particular radioactive material. While measures addressing the security functions of detection, delay and response should be included, the particular combination of measures could vary based on the situation specific analysis conducted when evaluating the security system. The performance based approach should consider the systematic interaction of detection, delay and response in determining overall system effectiveness against the assessed threat. Implementation of a performance based approach typically leads to a more tailored and cost effective set of security measures than is possible using the prescriptive approach.

6.68. Regulations calling for the use of a performance based approach should also include security management measures applicable to the security level of the radioactive material involved, as described in paras 6.36–6.63.

COMBINED APPROACH

6.69. States could also combine aspects of both the prescriptive and performance based approaches in order to apply security measures that meet the security goals and sub-goals for each security level for radioactive material. For example, a State could use the prescriptive approach for radioactive material with lower potential consequences of malicious use, but apply the performance based approach to the radioactive material of highest security concern. For such material, the operator would then be responsible for applying the appropriate security measures to meet a set of security sub-goals defined in terms of the security functions of detection, delay and response, as well as for the sub-goals related to security management.

Appendix I

DESCRIPTION OF SECURITY MEASURES

I.1.   Some of the security measures described below are referenced in Section 5. Others are intended to provide the reader with brief descriptions of additional measures which may be considered.

I.2.   Because national standards vary, this publication does not provide detailed guidance on specifications for security equipment or physical features. However, the design and reliability of security measures should be appropriate to the threat as identified by the national threat assessment or as defined in the DBT or RTS. Generally, this means the use of high quality, proven equipment and technology which satisfies national or international quality standards.

I.3.   The security measures are grouped according to the security functions of detection, delay and response. Security measures for security management are also addressed.

ACCESS CONTROL

I.4.   Access control can be exercised through entry checkpoints controlled by response personnel, the use of electronic readers or key control measures. Technology for access control, in the form of automatic access control systems (AACSs), is available in various forms, from simple pushbutton mechanical devices to more sophisticated readers that respond to proximity tokens or individual biometric characteristics. Used with a turnstile, AACS can also incorporate controls to inhibit practices such as 'pass back' and 'tailgating'. In most cases, the use of a card should be verified by a PIN keyed into the reader and in high security situations an AACS entry point should be supervised by a guard positioned within view.

I.5.   It is also important to limit access to the AACS management computers and software to prevent unauthorized modification of or interference with the system database.

CAGES

I.6.   Locked metal cages or containers can also be used to segregate and secure radioactive material by adding another level of protection (e.g. temporary retention within a receipt and dispatch area). Elsewhere, cages could be part of the storage arrangements within an established area that is enclosed and under control and supervision.

FENCES AND GATES

I.7.   The type of fence used on a perimeter should be appropriate to the threat, the nature of the radioactive material being protected and the category of the site overall. There are various types of fence, ranging from those that are little more than a demarcation to those that are more robust and can be combined with a fence mounted perimeter intrusion detection and assessment system or electrified panels. Fence lines need to be checked regularly to ensure that the fabric is in good order and free from interference or damage. Gates within a fence should be constructed to a standard comparable to or higher than that of the fence and secured with good quality locks.

INTRUSION DETECTION SYSTEMS

I.8.   Intrusion detection systems are a useful means of monitoring the security of an unoccupied area. Where appropriate, the technology can be extended to the outer area of an establishment by use of a perimeter intrusion detection and assessment system (with fence vibration sensors, external motion sensors, infrared and microwave detectors, underground step sensors). Intrusion detection systems can be supplemented by sensors to detect vibrations and the opening of doors or windows, breaking or cutting of glass and dismantling of walls. All intrusion detection systems should be supported by response measures to investigate alarm events or conditions. Alarms can sound remotely at a security control point, locally through a high volume sounder or both. Video monitoring can be a useful aid in providing initial verification of events within an alarmed zone or area but should normally be backed up by a patrol making a visual check or investigation.

KEY CONTROL PROCEDURES

I.9.   Keys which allow access to radioactive material should be controlled and secured. These could be keys to cages, doors, storage containers or shielded units within which radioactive material is used. Similar levels of control should be applied to duplicate and spare keys.

LOCKS, HINGES AND INTERLOCKS FOR DOORS

I.10. Locks used for the protection of radioactive material should be of good quality, incorporating features that will offer some resistance to forcible attack. The same applies to hinges on doors. Keys should be safeguarded in the manner outlined in the measures described for security management. Within premises, interlock doors that meet safety requirements can serve the interests of security by controlling the movement of personnel and allowing staff to monitor access to the facility. Where conventional locks and keys are used as a means of control, locks should be of good quality and key management procedures should be designed to prevent unauthorized access or compromise.

LOCKED, SHIELDED CONTAINERS

I.11. Shielding and fixed units containing radioactive material can provide protection, and can delay any attempt to interfere with that material. However, when operator personnel are not present, the area should be covered by an intruder detection alarm system to alert the response personnel or security response of the need to investigate the circumstances of any intrusion.

QUALITY ASSURANCE

I.12. Security arrangements and procedures should be prepared, documented and maintained in line with recommended quality assurance standards such as recording of formal approval; version control; periodic, planned review; testing of arrangements and procedures; and incorporation of lessons identified into procedures.

STANDBY POWER

I.13. Security control rooms and security systems should be able to cope with power dips or outright loss of a main electricity supply. This can be ensured through an uninterruptible power supply and a standby generator which automatically starts when a fluctuation in power levels is detected. Battery backup has only limited duration and should therefore be viewed as a short term source of standby power.

TWO PERSON RULE

I.14. Certain areas can only be accessed by at least two persons at the same time.

VIDEO MONITORING

I.15. Video monitoring is a useful aid which allows security staff to monitor outer approaches and areas where radioactive material is stored. Cameras can be combined with an intrusion detection system to provide event activated camera views along with video capture to allow assessment of an alarm even though the cause of the alarm may no longer be in the immediate vicinity. However, to be fully effective, video cameras and monitors should be regularly assessed to ensure that they continue to display imagery of good quality. Systems should also be supported by a response so that alarm events and indications activated by technology can be investigated. The whole video surveillance and assessment system can consist of analogue and digital (IP based) cameras, infra reflectors, coaxial and bunched conductor pairs, optical and wireless image transmission devices and monitors.

WALLS

I.16. Walls can provide effective protection against unauthorized access to a facility. However, unless they are already in place, walls are an expensive way to form a perimeter boundary.

WINDOWS AND DOORS

I.17. Windows and doors should exhibit sufficient penetration resistance against an intruder. The windows should comply with the same requirements as the doors, which could be ensured by security glass or by a fixed security grill that cannot be disassembled from outside or by an inside security grill that can be opened, is fully welded and is made of appropriate steel. The window and door casings and frames should exhibit at least the same resistance as the door and the glass.

Appendix II

TOPICS TO BE ADDRESSED IN AN OPERATOR'S SECURITY PLAN

II.1. The purpose of a security plan is to describe the security system and procedures that are in place to protect radioactive material in use and in storage and associated facilities. The following annotated outline provides high level guidance for drafting a security plan, including suggested topics and content that should be considered within each topic. Certain sections of the security plan could be developed separately (e.g. the response plan), but should be referenced in the security plan consistent with information security requirements.

1.   INTRODUCTION

Objective(s) of the security plan

Describe the objectives to be satisfied by the security plan, such as documenting the operation of the security system and security management measures in order to meet or demonstrate compliance with regulatory requirements.

Scope

Briefly describe the areas to be covered by the security plan, including the plan's link to other relevant documents or arrangements such as any management system, operational safety, radiation protection or emergency preparedness and response matters.

Preparation and updating

Describe the process for developing, updating and approving the security plan.

2.   FACILITY DESCRIPTION

This section should describe the radioactive material(s) and their location(s); the level of protection required according to the categorization of the material and the assessed security level; the physical features of the facility; and the facility's operations and regulatory requirements.

3.   SECURITY MANAGEMENT

This section should describe the security management measures in place, including:

— Roles and responsibilities;
— Training and qualification;
— Access authorization;
— Trustworthiness;
— Information protection;
— Maintenance programme;
— Budget and resource planning;
— Evaluation for compliance and effectiveness.

4.   SECURITY SYSTEM

This section should describe how the security system achieves the required level of protection, based on a graded approach. The specific measures to be described should include the following.

Threat information

To the extent that the threat information is provided by the regulatory body, describe the information in sufficient detail to indicate how the security system is designed to protect against both external and internal threats. Also indicate who is responsible for receiving threat information and how such information is shared with operator personnel who have a need to know.

Security assessment methodology

Describe the process or methodology used to evaluate the security system and assess its vulnerabilities, taking into account the threat information provided.

Security system design

Describe how the security system has been designed to provide the level of protection required, taking into account the graded approach and the principles of defence in depth and balanced protection. This section should also describe modifications to the security system in the case of increased threat.

Access control

Describe the physical measures for controlling access, including how personnel and vehicles are physically controlled at each access control point to limit access only to authorized persons and the specific media used to authenticate the identity of authorized persons and vehicles at access points such as key card, personal identification number, biometric device or a combination thereof.

Delay, detection and alarm assessment measures

For each of the controlled or secured areas, describe the means of detection at each barrier or access point, the barriers (delay measures) used to increase adversary task time relative to response time and the methods of alarm assessment (such as video monitoring, central alarm stations, both internal and external guard or response forces, and computer and recording systems).

5. SECURITY PROCEDURES

This section should describe the written procedures for personnel, such as procedures for routine, off-shift and emergency operations, opening and closing of the facility, key and lock control, accounting and inventory control, and acceptance and transfer of the radioactive material from one facility to another.

6. RESPONSE

This section should describe the response arrangements for all nuclear security events, including references to emergency plans and emergency response actions. This section should capture the following:

— Roles and responsibilities of on-site security or facility personnel during nuclear security events and those of local and national response forces if external response is required;
— Communication methods to be used by response forces when communicating with the alarm monitoring station or facility security personnel;
— Procedures for reporting nuclear security events, including any reporting requirements and arrangements for review of the security system following an event and corrective actions required.

REFERENCES

List any reference documents, such as specific regulations, regulatory authorization, operating manuals, organizational policies and manuals, that are referred to in the security plan or are needed to explain or expand on any details in the plan.

Appendix III

DESCRIPTION OF A VULNERABILITY ASSESSMENT

III.1. There are a number of methods that can be used to verify that facilities are in compliance with all applicable security requirements and to assess the effectiveness of their security systems. One such method is a vulnerability assessment, a method of evaluating the effectiveness of a facility's security system.

III.2. Examples of vulnerabilities within a facility include:

— Ineffective or absent security measures;
— Inappropriate administrative controls;
— Inadequate communication;
— Poor security culture;
— Incompatibility of security measures with safety measures.

III.3. Vulnerability should be assessed against the basic functions of security (detection, delay and response) and security management to ensure that the risks associated with malicious acts against radioactive material and associated facilities, as defined by the State, are managed to an acceptable level.

III.4. A vulnerability assessment is a systematic appraisal of the effectiveness of a security system in protecting against a threat. The vulnerability assessment can be specific or general in nature. It can be conducted locally by the operator to demonstrate system effectiveness against the requirements specified by the regulatory body, or to design or make modifications to the existing design of the security system. The vulnerability assessment can also be conducted and used by the regulatory body in developing or evaluating either its regulations or the operator's security system.

III.5. Those conducting the vulnerability assessment should be technical experts familiar with the facility in question, particularly its technical and commercial operations, with the appropriate knowledge and skills related to the design and evaluation of security systems.

III.6. The VA process comprises three major phases:

— *Planning the vulnerability assessment* includes determining the scope and objectives of the VA; selecting a methodology; evaluating potential threats and their capabilities; understanding the nature of the facility, including the attractiveness of the material and the threat environment; defining the roles and responsibilities of the vulnerability assessment team; determining the resources and time frame required to complete the assessment; confirming the radioactive material inventory and associated information; and taking note of the categorization, form and location of the radioactive material and the physical environment in which it is located.

— *Conducting the vulnerability assessment* includes defining the requirements of the security system; gathering the data needed to characterize the security system and its components; analysing the ability of the system to meet the requirements; identifying existing security measures; assessing the expected effectiveness of the security system in protecting against attacks by the assessed threats; and determining what, if any, additional security measures are necessary to meet the required level of protection.

— *Completing the vulnerability assessment* includes the provision of reports outlining the methodology used, the assumptions made, the data collected, the effectiveness of the security system and recommendations for upgrades, if required.

REFERENCES

[1]   INTERNATIONAL ATOMIC ENERGY AGENCY, Objective and Essential Elements of a State's Nuclear Security Regime, IAEA Nuclear Security Series No. 20, IAEA, Vienna (2013).

[2]   INTERNATIONAL ATOMIC ENERGY AGENCY, Nuclear Security Recommendations on Physical Protection of Nuclear Material and Nuclear Facilities (INFCIRC/225/Revision 5), IAEA Nuclear Security Series No. 13, IAEA, Vienna (2011).

[3]   INTERNATIONAL ATOMIC ENERGY AGENCY, Nuclear Security Recommendations on Radioactive Material and Associated Facilities, IAEA Nuclear Security Series No. 14, IAEA, Vienna (2011).

[4]   EUROPEAN POLICE OFFICE, INTERNATIONAL ATOMIC ENERGY AGENCY, INTERNATIONAL CIVIL AVIATION ORGANIZATION, INTERNATIONAL CRIMINAL POLICE ORGANIZATION–INTERPOL, UNITED NATIONS INTERREGIONAL CRIME AND JUSTICE RESEARCH INSTITUTE, UNITED NATIONS OFFICE ON DRUGS AND CRIME, WORLD CUSTOMS ORGANIZATION, Nuclear Security Recommendations on Nuclear and Other Radioactive Material out of Regulatory Control, IAEA Nuclear Security Series No. 15, IAEA, Vienna (2011).

[5]   INTERNATIONAL ATOMIC ENERGY AGENCY, Code of Conduct on the Safety and Security of Radioactive Sources, IAEA/CODEOC/2004, IAEA, Vienna (2004).

[6]   International Convention for the Suppression of Acts of Nuclear Terrorism, United Nations, New York (2005).

[7]   INTERNATIONAL ATOMIC ENERGY AGENCY, Guidance on the Management of Disused Radioactive Sources, IAEA/CODEOC/MGT-DRS/2018, IAEA, Vienna (2018).

[8]   INTERNATIONAL ATOMIC ENERGY AGENCY, Guidance on the Import and Export of Radioactive Sources, IAEA/CODEOC/IMO-EXP/2012, IAEA, Vienna (2012).

[9]   FOOD AND AGRICULTURE ORGANIZATION OF THE UNITED NATIONS, INTERNATIONAL ATOMIC ENERGY AGENCY, INTERNATIONAL CIVIL AVIATION ORGANIZATION, INTERNATIONAL LABOUR ORGANIZATION, INTERNATIONAL MARITIME ORGANIZATION, INTERPOL, OECD NUCLEAR ENERGY AGENCY, PAN AMERICAN HEALTH ORGANIZATION, PREPARATORY COMMISSION FOR THE COMPREHENSIVE NUCLEAR-TEST-BAN TREATY ORGANIZATION, UNITED NATIONS ENVIRONMENT PROGRAMME, UNITED NATIONS OFFICE FOR THE COORDINATION OF HUMANITARIAN AFFAIRS, WORLD HEALTH ORGANIZATION, WORLD METEOROLOGICAL ORGANIZATION, Preparedness and Response for a Nuclear or Radiological Emergency, IAEA Safety Standards Series No. GSR Part 7, IAEA, Vienna (2015).

[10] FOOD AND AGRICULTURE ORGANIZATION OF THE UNITED NATIONS, INTERNATIONAL ATOMIC ENERGY AGENCY, INTERNATIONAL LABOUR OFFICE, PAN AMERICAN HEALTH ORGANIZATION, UNITED NATIONS OFFICE FOR THE COORDINATION OF HUMANITARIAN AFFAIRS, WORLD HEALTH ORGANIZATION, Arrangements for Preparedness for a Nuclear or Radiological Emergency, IAEA Safety Standards Series No. GS-G-2.1, IAEA, Vienna (2007).

[11] INTERNATIONAL ATOMIC ENERGY AGENCY, Security of Radioactive Material in Transport, IAEA Nuclear Security Series No. 9-G, IAEA, Vienna (in preparation).

[12] INTERNATIONAL ATOMIC ENERGY AGENCY, Physical Protection of Nuclear Material and Nuclear Facilities (Implementation of INFCIRC/225/Revision 5), IAEA Nuclear Security Series No. 27-G, IAEA, Vienna (2018).

[13] STOIBER, C., BAER, A., PELZER, N., TONHAUSER, W., Handbook on Nuclear Law, IAEA, Vienna (2003).

[14] STOIBER, C., CHERF, A., TONHAUSER, W., VEZ CARMONA, M.L., Handbook on Nuclear Law: Implementing Legislation, IAEA, Vienna (2010).

[15] INTERNATIONAL ATOMIC ENERGY AGENCY, Governmental, Legal and Regulatory Framework for Safety, IAEA Safety Standards No. GSR Part 1 (Rev. 1), IAEA, Vienna (2016).

[16] EUROPEAN COMMISSION, FOOD AND AGRICULTURE ORGANIZATION OF THE UNITED NATIONS, INTERNATIONAL ATOMIC ENERGY AGENCY, INTERNATIONAL LABOUR ORGANIZATION, OECD NUCLEAR ENERGY AGENCY, PAN AMERICAN HEALTH ORGANIZATION, UNITED NATIONS ENVIRONMENT PROGRAMME, WORLD HEALTH ORGANIZATION, Radiation Protection and Safety of Radiation Sources: International Basic Safety Standards, IAEA Safety Standards Series No. GSR Part 3, IAEA, Vienna (2014).

[17] INTERNATIONAL ATOMIC ENERGY AGENCY, Developing Regulations and Associated Administrative Measures for Nuclear Security, IAEA Nuclear Security Series No. 29-G, IAEA, Vienna (2018).

[18] INTERNATIONAL ATOMIC ENERGY AGENCY, Security of Nuclear Information, IAEA Nuclear Security Series No. 23-G, IAEA, Vienna (2015).

[19] EUROPEAN POLICE OFFICE, INTERNATIONAL ATOMIC ENERGY AGENCY, INTERNATIONAL POLICE ORGANIZATION, WORLD CUSTOMS ORGANIZATION, Combating Illicit Trafficking in Nuclear and Other Radioactive Material, IAEA Nuclear Security Series No. 6, IAEA, Vienna (2007).

[20] INTERNATIONAL ATOMIC ENERGY AGENCY, Operations Manual for Incident and Emergency Communication, EPR-IEComm 2012, IAEA, Vienna (2012).

[21] INTERNATIONAL ATOMIC ENERGY AGENCY, IAEA Response and Assistance Network, EPR-RANET 2018, IAEA, Vienna (2018).

[22] INTERNATIONAL ATOMIC ENERGY AGENCY, Nuclear Security Systems and Measures for the Detection of Nuclear and Other Radioactive Material out of Regulatory Control, IAEA Nuclear Security Series No. 21, IAEA, Vienna (2013).

[23] INTERNATIONAL ATOMIC ENERGY AGENCY, Development, Use and Maintenance of the Design Basis Threat, IAEA Nuclear Security Series No. 10, IAEA, Vienna (2009).

[24] INTERNATIONAL ATOMIC ENERGY AGENCY Preventive and Protective Measures against Insider Threats, IAEA Nuclear Security Series No. 8, IAEA, Vienna (2008).

[25] INTERNATIONAL ATOMIC ENERGY AGENCY Enhancing Nuclear Security Culture in Organizations Associated with Nuclear and Other Radioactive Material, IAEA Nuclear Security Series, IAEA, Vienna (in preparation).

[26] FOOD AND AGRICULTURE ORGANIZATION OF THE UNITED NATIONS, INTERNATIONAL ATOMIC ENERGY AGENCY, INTERNATIONAL LABOUR OFFICE, PAN AMERICAN HEALTH ORGANIZATION, WORLD HEALTH ORGANIZATION, Criteria for Use in Preparedness and Response for a Nuclear or Radiological Emergency, IAEA Safety Standards Series No. GSG-2, IAEA, Vienna (2011).

[27] INTERNATIONAL ATOMIC ENERGY AGENCY, Sustaining a Nuclear Security Regime, IAEA Nuclear Security Series No. 30-G, IAEA, Vienna (2018).

[28] INTERNATIONAL ATOMIC ENERGY AGENCY, Nuclear Security Culture, IAEA Nuclear Security Series No. 7, IAEA, Vienna (2008).

[29] INTERNATIONAL ATOMIC ENERGY AGENCY, Categorization of Radioactive Sources, IAEA Safety Standards Series No. RS-G-1.9, IAEA, Vienna (2005).

[30] INTERNATIONAL ATOMIC ENERGY AGENCY, Dangerous Quantities of Radioactive Material (D-Values), EPR-D-VALUES 2006, IAEA, Vienna (2006).

[31] INTERNATIONAL ATOMIC ENERGY AGENCY, Classification of Radioactive Waste, IAEA Safety Standards Series No. GSG-1, IAEA, Vienna (2009).

IAEA
International Atomic Energy Agency

ORDERING LOCALLY

IAEA priced publications may be purchased from the sources listed below or from major local booksellers.

Orders for unpriced publications should be made directly to the IAEA. The contact details are given at the end of this list.

NORTH AMERICA

Bernan / Rowman & Littlefield
15250 NBN Way, Blue Ridge Summit, PA 17214, USA
Telephone: +1 800 462 6420 • Fax: +1 800 338 4550
Email: orders@rowman.com • Web site: www.rowman.com/bernan

Renouf Publishing Co. Ltd
22-1010 Polytek Street, Ottawa, ON K1J 9J1, CANADA
Telephone: +1 613 745 2665 • Fax: +1 613 745 7660
Email: orders@renoufbooks.com • Web site: www.renoufbooks.com

REST OF WORLD

Please contact your preferred local supplier, or our lead distributor:

Eurospan Group
Gray's Inn House
127 Clerkenwell Road
London EC1R 5DB
United Kingdom

Trade orders and enquiries:
Telephone: +44 (0)176 760 4972 • Fax: +44 (0)176 760 1640
Email: eurospan@turpin-distribution.com

Individual orders:
www.eurospanbookstore.com/iaea

For further information:
Telephone: +44 (0)207 240 0856 • Fax: +44 (0)207 379 0609
Email: info@eurospangroup.com • Web site: www.eurospangroup.com

Orders for both priced and unpriced publications may be addressed directly to:

Marketing and Sales Unit
International Atomic Energy Agency
Vienna International Centre, PO Box 100, 1400 Vienna, Austria
Telephone: +43 1 2600 22529 or 22530 • Fax: +43 1 26007 22529
Email: sales.publications@iaea.org • Web site: www.iaea.org/publications